SpringerBriefs in Computer Science

SpringerBriefs present concise summaries of cutting-edge research and practical applications across a wide spectrum of fields. Featuring compact volumes of 50 to 125 pages, the series covers a range of content from professional to academic.

Typical topics might include:

- A timely report of state-of-the art analytical techniques
- A bridge between new research results, as published in journal articles, and a contextual literature review
- A snapshot of a hot or emerging topic
- An in-depth case study or clinical example
- A presentation of core concepts that students must understand in order to make independent contributions

Briefs allow authors to present their ideas and readers to absorb them with minimal time investment. Briefs will be published as part of Springer's eBook collection, with millions of users worldwide. In addition, Briefs will be available for individual print and electronic purchase. Briefs are characterized by fast, global electronic dissemination, standard publishing contracts, easy-to-use manuscript preparation and formatting guidelines, and expedited production schedules. We aim for publication 8–12 weeks after acceptance. Both solicited and unsolicited manuscripts are considered for publication in this series.

More information about this series at http://www.springer.com/series/10028

Deze Zeng • Lin Gu • Shengli Pan • Song Guo

Software Defined Systems

Sensing, Communication and Computation

 Springer

Deze Zeng
School of Computer Science
China University of Geosciences
Wuhan, Hubei, China

Shengli Pan
School of Computer Science
China University of Geosciences
Wuhan, Hubei, China

Lin Gu
Huazhong University of Science
and Technology
Wuhan, Hubei, China

Song Guo
Department of Computing
The Hong Kong Polytechnic University
Hong Kong, Kowloon, Hong Kong

ISSN 2191-5768   ISSN 2191-5776 (electronic)
SpringerBriefs in Computer Science
ISBN 978-3-030-32941-9   ISBN 978-3-030-32942-6 (eBook)
https://doi.org/10.1007/978-3-030-32942-6

This Springer imprint is published by the registered company Springer Nature Switzerland AG.
The registered company address is: Gewerbestrasse 11, 6330 Cham, Switzerland

Dedicated to our family, friends, colleagues, and to all who are interested in software defined systems.

Preface

The heavy reliance on hardware hinders the innovation in information systems ranging from the most front-end devices to the most back-end servers. It has been shown as a compelling trend towards software defined system (SDS), where all resources can be managed in a software-defined way. Thanks to the recent development in information technologies, many enabling technologies for SDS are available already.

In this book, we first introduce the SDS concept, architecture, and its enabling technologies such as software defined sensor networks (SDSN), software defined radio, cloud/fog radio access networks (C/F-RAN), software defined networking (SDN), network function virtualization (NFV), software defined storage, virtualization, and docker. We then, respectively, discuss the resource allocation and task scheduling in SDS, mainly focusing on sensing, communication, networking, and computation. Related case studies on SDSN, C/F-RAN, SDN, and NFV will be given, and we also discuss how these technologies cooperate with each other to enable cross resource management and task scheduling in SDS. Novel resource allocation and task scheduling algorithms will be introduced and evaluated.

The intended audience of this book shall be the readers who are interested in the cutting-edge information system softwarization technologies, resource allocation and optimization algorithm design, performance evaluation and analysis, next-generation communication and networking technologies, edge computing, cloud computing, and IoT.

Wuhan, China   Deze Zeng
Wuhan, China   Lin Gu
Wuhan, China   Shengli Pan
Hong Kong, Hong Kong   Song Guo
August 2019

Acknowledgements

This work was partially supported by the National Natural Science Foundation of China (under Grant No. 61701074, 61772480, 61402425, and 61673354), the Fundamental Research Funds for the Central Universities, China University of Geosciences (Wuhan) (under Grant No. G1323519020), and the Open Research Project of The Hubei Key Laboratory of Intelligent Geo-Information Processing.

Contents

Acronyms

AC	Access Control
AI	Artificial Intelligence
BBU	BaseBand Unit
BPOS	Backpressure Based Online Scheduling
CapEx	Capital Expenditures
CDF	Cumulative Distribution Function
CoMP	Coordinated Multi-Point
COTS	Commercial-Off-The-Shelf
CR	Cognitive Radio
CRAN	Cloud-Radio Access Networks
CSI	Channel State Information
DER	Distributed Renewable Resources
DPI	Deep Packet Inspection
FRAN	Fog Radio Access Networks
IaaS	Infrastructure as a Service
ICT	Information and Communications Technology
ILP	Integer Linear Programming
IoT	Internet of Things
LP	Linear Programming
MC-SCA	Minimum Cost Switch-Controller Association
MDP	Markov Decision Process
MEC	Mobile Edge Computing
MILP	Mixed-Integer Linear Programming
MIQP	Mixed-Integer with Quadratic constraints Programming
ML	Machine Learning
MPR	Multi-Path Routing
NAT	Network Address Translation
NF	Network Function
NFV	Network Function Virtualization
ONF	Open Networking Foundation
OpEx	Operating Expenditures

OSS/BSS	Operation Support System/Business Support System
PaaS	Platform as a Service
PoI	Point-of-Interest
QIP	Quadratic Integer Programming
QoE	Quality of Experience
QoS	Quality of Service
RF	Radio Frequency
RL	Reinforcement Learning
RRH	Remote Radio Head
SaaS	Software as a Service
SC	Service Chaining
SDN	Software Defined Network
SDS	Software Defined System
SDSN	Software Defined Sensor Network
SHM	Structural Health Monitoring
SINR	Signal-to-Interference-plus-Noise Ratio
SOA	Service-Oriented Architecture
SOC	Second-Order Cone
SPR	Shortest-Path Routing
TCAM	Ternary Content Addressable Memory
UE	User Equipment
UI	User Interface
VM	Virtual Machines
VNF	Virtualized Network Function
WAN	Wireless Area Network
WSN	Wireless Sensor Network
XaaS	Everything as a Service

Chapter 1
Introduction

Abstract The newly emerged software defined system (SDS) promises a new information system resource allocation and management way. The main concept of SDS is to separate the control plane from the data plane in various subsystems, e.g., sensing, communication, networking, and computation. By such means, various resources of the information system are virtualized and therefore can be managed in a more friendly and flexible manner. It is widely believed that SDS is able to lower the barrier for system and application innovation, and will become an inevitable trend towards the future generation of the information system. In this chapter, we first identify the emergence and then give an overview as well as the main concepts of SDS. Regarding that many enabling technologies are already available to realize the vision of SDS, we also present a brief summarization of the main enabling technologies for SDS and identify their key characteristics.

1.1 The Rising of Software Defined Systems

Noticed that conventional information systems get the benefit of easy management due to the tight binding with hardware. However, using dedicated equipment to meet challenges might be effective but very costly. It is widely recognized that future information system like SDS shall be a networked system as the sound convergence of various resources including sensing, communication, networking, and computation. SDS, which aims at controlling virtual resource allocation through a centralized control plane, can proactively configure network resources for services. It can not only mitigate the high cost, but also greatly improve the automation of the information system. However, achieving such SDSs could be very complex. Their control and management platforms of conventional information systems are facing considerable challenges regarding flexibility, dependability, and scalability. As a result, the next generation of the information systems will require a paradigm shift in how they are constructed and managed. To break such hard reliance, "softwarizing" the information resources has become a compelling, and inevitable, development trend.

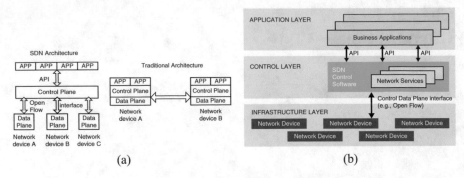

Fig. 1.1 Illustrations of SDN. (**a**) SDN vs. traditional network architecture. (**b**) Layers of SDN

For example, the newly emerged networking paradigm, i.e., "software defined network" (SDN), has shown a great potential in overcoming the above limitations of conventional computer networks [56]. As illustrated in Fig. 1.1a, SDNs provide network agility, programmability, and most importantly, the centralized network control rather than the distributed control in traditional networks. These new features facilitate the management, performance, and QoS guarantee in the next-generation networks [31]. It is a popular way to employ SDN cloud/edge computing paradigm to alleviate some of the problems associated with resource allocation, utilization and management of today's increasingly complex networks. Thereout, many customized and innovative network applications therefore can be developed by programmers or network administrators.

Although so, many elements of such networks, which could refer to the underlying networking topologies, many aspects of the user control over "Infrastructure-as-a-Service" (IaaS), "Platform-as-a-Service" (PaaS), and "Software-as-a-Service" (SaaS) layers, the construction of "Everything-as-a-Service" (XaaS) services, provenance, and meta-data collection [44], etc., still remain "stiff" and hard to modify, lacking of the capability to adapt in an integrated fashion [27]. It has been noticed that in many pioneering solutions for SDS, their service abstractions seem to be selective and inadequate. To achieve a more comprehensive service abstraction, the SDS concept therefore will not only just include networking, but also should take a comprehensive and joint consideration of sensing, communication and computation.

To this end, softwarization has become the ideal facilitator for a new generation of highly flexible, agile network systems, leading to innovative new products and a more standard, simplified infrastructure. Some key capabilities for SDS according to the Open Networking Foundation (ONF) description are summarized in the following Table 1.1. Based on these capabilities, SDSs are able to span vendors and devices, and facilitate the dynamic management of the services with a new level of automation. These will not only allow managers to re-focus from managing bytes and packets to managing the overall "Quality-of-Experience" (QoE), but also support them to achieve rapid service delivery and elastic capacity required for

Table 1.1 Key capabilities of SDS

Characteristics	Description
Agility	Network managers can dynamically adjust network-wide traffic flows to meet changing needs without having to separately re-configure routers
Programmability	Network control functions are easier to program as a result of both the centralization and the separation of control from processing
Manageability	Network intelligence in the (logically) centralized SDS controllers can maintain a global view of the network and can appear to applications and policy engines as a single, logical switch
Configurability	Network managers can more easily and more dynamically configure, manage, secure, and optimize network resources
Interoperability	An SDS using open standards can simplify network design and operation since forwarding instructions are provided by the SDS controllers instead of multiple, vendor-specific devices and protocols
Protectability	Robust security and privacy policies and controls can be applied to SDS network elements in an automated and consistent network-wide fashion

cloud/edge environments [99]. As illustrated in Fig. 1.1b, the essence of the SDN architecture is to well divide the network into three subsystems (usually called layers):

- **Application Layer**: provides user applications with programmatic interfaces to the network services regardless of the underlying resources; OpenStack is one implementation of such an interface [84];
- **Control Layer**: provides a separate programmable facility for the control and management of the network infrastructure;
- **Infrastructure Layer**: provides open standards-based interfaces to the basic IT resources (i.e., networks, computing, and storage); OpenFlow is an example [44].

As service providers accelerate the trial and adoption of SDS solutions, they are increasingly realizing the limitations and barriers of existing operation support system/business support system (OSS/BSS) systems. A key problem facing service providers/operators is how to build on the promises of SDN, without numerous risks and complexities of overhauling massive legacy OSS/BSSs that currently support revenue-generating services. For most service providers, this is a question of intelligent evolution versus potentially costly revolution. Network function virtualization (NFV) is an efficient way that complements SDN. It disunites software from hardware to enable flexible network deployment and dynamic operation, and uses commodity servers to run network services software versions that previously were hardware-based [102]. However, this is not everything. Increasing network complexity directly relates to the increase in operational expenditure to run and manage the networks. Besides, the complexity of today's networks is also extensive, and it will only increase in the coming years. Exploring how we are using artificial intelligence (AI), such as the machine learning methods, to enhance the performance and to augment human capabilities to improve the efficiency [69], will be a promising way to address the increasing complexities of the network systems nowadays.

1.2   Overview of Software Defined Systems

Nowadays, it is moving into a new era where the information and communications technology (ICT) is "software-defined" and typically offered "as a Service", e.g., the well-known SaaS, PaaS, and IaaS. This is especially true for enterprise networks although the concept can be extended to computing, storage and transmission as well. An SDS enables central, self-managing control over the full range of computing resources: processing, storage, network, security, edge and external cloud. The SDS model intelligently links the control interface of individual devices into a dynamic, orchestrated system. This results in a real-time optimized pipeline in which overlapping workloads run efficiently, reliably, and at a lower yet practical cost. This could be generalized to the physical/virtual separation of the control subsystem from the functional (processing/storage/networking/device) subsystems, while the control subsystem controls the functional subsystems.

The key inspiration of SDS lies in SDN, which attempts to centralize network intelligence in one network component by disassociating the forwarding process of network packets (data plane) from the routing process (control plane) [44]. The control plane consists of one or more controllers which are considered as the brain of SDN network where the whole intelligence is incorporated. Since the latter's emergence in 2011, SDN was commonly associated with the OpenFlow protocol (for remote communication with network plane elements for the purpose of determining the path of network packets across network switches). This builds the data plane, also known as forwarding plane. It forwards traffic to the next hop along the path to the selected destination network according to control plane logic. The data plane packets go through the router, while the routers/switches use the policies built by the control plane to process these data plane packets.

Thanks to the splitting of the control and data forwarding functions, the architecture of SDN gives the applications more information about the state of the entire network from the controller, as opposed to traditional networks. SDS borrows the same concept to separate and centralize the control plane. Beyond network control, SDS enriches the resources and elements in centralized control. The architecture APIs of SDN are often referred to as northbound and southbound interfaces, defining the communication between the applications, controllers, and networking systems. SDS inherits the same architecture. A northbound interface is defined as the connection between the controller and applications, whereas the southbound interface is the connection between the controller and the physical hardware.

One of the key strengths of the SDS architecture is that it provides applications with the unique ability to obtain an abstracted view of the entire information system. These make the system "smarter" by being able to analyze itself and integrate real-time information about system activity according the application requirements. To this end, the practical implementation of SDS requires the comprehensively central control capability of sensing, communication, networking, and computation.

- Sensing fills the gap between the physical world and the cyber world. Internet of things (IoT) has been widely deployed to bring the physical world information into the cyber world. Traditional IoT applications were usually built on application-specific sensor networks, whose sensing behaviors were built in and hard to be altered. Software defined sensing is therefore proposed to softwarize the sensor networks, whose sensing behaviors can be centrally controlled.
- Communication, especially the one in wireless access networks, plays a critical role in the information system. In traditional wireless networks, all the wireless communication resources, e.g., power, frequency, were all controlled in a distributed manner by respective base stations. Recently, a new concept named as cloud-radio access networks (CRAN), or fog radio access networks (FRAN), enables a new wireless access network paradigm, where it is possible to orchestrate the wireless communication behavior in a centralized manner.
- Network was the first one that attracts the attention of researchers and engineers to pursue centralized control. Actually, we think SDN is the key and base inspiration of SDS. SDN focuses on the central control of network elements while SDS broaden the resources and elements to control, so as to enable a more comprehensive centrally controllable system.
- Computation is another key aspect requiring flexibility. To make a unified software defined system, besides sensing, communication and network, we also hope that computation could also be software-definable. Various virtualization technologies, e.g., VM, container, have hide the underline physical heterogeneity to provide a unified interface for the manipulation of computation resources. Both cloud computing and edge computing actually build upon such software-definability.

Taking in all above issues, SDS promises a new running paradigm of information systems that all the elements and resources can be centrally controlled and programmable. Unlike traditional information systems where we have to carefully tune our applications according to the characteristics built in individual subsystems, SDS programmers, or system administrators, could manage the system behaviors flexibly and intelligently.

1.3 Enabling Technologies

1.3.1 Software Defined Network

SDN was first proposed in 2006 by the Clean Slate project at Stanford University. The purpose was to redesign the Internet, whose design at that time was slightly out of date and stiff to evolve [25]. Their practical goal was to achieve an implementation of network virtualization. The emergence of SDN and NFV [75] as two pillars of a new networking paradigm has led to design and deployment of elastic content- and service-delivery virtual networks on the new types of networks, especially the cloud-based data centers whose backbones consist of software-defined substrate physical networks. This dynamic, multi-layer architecture for network functions introduces new challenges in maintaining performance and quality of service for service providers, and inspires new ideas and frameworks in performance measurement and network management [26].

The core technology of SDN is OpenFlow, which realizes flexible control of network traffic by separating the control plane of the network device from the data plane [58]. The core appeal of SDN is to simplify network operation and maintenance through automated service deployment. It will be a difficult task to meet this demand without separating control and forwarding. But, noticed that the separation of control and forwarding is only a means to satisfy the core appeal of SDN. There are also other means in some scenarios, such as separation of management and control. Nevertheless, one should not achieve the design concept of SDN at the cost of significantly increasing the networking complexity [6]. To achieve flexible control of network traffic and build the network more intelligent as a pipeline, the network needs to be abstracted to shield the underlying complexity and to only provide simple and efficient configuration and management for the upper layer. It is widely recognized that SDN will settle a good platform for innovation in core networks and applications. Here, in order to further illustrate the advantage of SDN, some disadvantages of the conventional networks are outlined as follows:

– The flow increases unexpectedly, and the existing network is difficult to adapt to the demand for massive data transmission;
– The network structure is single and not flexible enough [15]. The previous superposition mode cannot be realized and adapts to the emerging new services;
– Not only the problem of network sustainable development becomes more and more serious, but also other problems such as network security and network uncontrollability [6, 95] seem to more frequently appear;
– Technologies such as cloud computing and big data [38] have proposed new network feature requirements, which incur the imbalance of the complexity and cost concerns to existing networks.

Accordingly, if the traditional networks evolve forward to meet all the challenges, it will be necessary to re-edit the corresponding network equipment (routers, switches, firewalls [6] and etc.), which would be very complicated to achieve. Noticed that, once any serious network abnormal/bug occurs, it might affect the whole network operation and will be very difficult to troubleshoot. Moreover, the Internet environment changes more rapidly nowadays than ever. The requirements for flexibility and convenience are higher when the security and stability of the network are not guaranteed. Consequently, SDN will be a key option which can exactly separate the control on the network equipment and manage them with the centralized controller. The control (rights), which do not need to rely on the underlying network equipment and transparently handle the differences from the underlying network equipment, are totally open [22]. It is then possible for users/operators to change any network routing and transport rule policies as needed. As you can see, to implement

SDN needs to subvert existing networks by means of transfer separation, centralized control, and open programmable [97]. This could be summarized as the following options:

- **Open Protocol:** Such a protocol scheme creates an ideal network architecture according to the SDN concept, and can efficiently separate the control layer and the forwarding layer [58]. It is the most revolutionary solution that can be launched for the user to get rid of the vendor lock-in. To achieve this, it can refer to protocols such as ONF SDN and ETSI NFV. Nevertheless, this network program/development requirements might be also higher than ever. Currently, there are only a handful of vendors that can afford, like Huawei, Brocade, Dell, and NCIRA, etc.
- **Overlay Network:** This solution realizes network resource pooling by creating a virtual network based on the original network to isolate the difference and complexity between the underlying devices [25]. Logical separation of existing network resources and management of the network using a multi-tenant model to better meet the needs of emerging services such as big data and cloud computing. At present, the main implementation schemes include VXLAN, NVGRE, NVP, etc.
- **Dedicated Interface:** This implementation idea is different from the above two. It does not change the implementation mechanism and working mode of the traditional network, but develops a dedicated API interface on the network device by changing the network device and the operating system (OS) [5]. The administrator can implement the unified configuration management and delivery of the network device through the API interface, replacing the manual operation mode that requires the login configuration of one device. At the same time, these API interfaces are also available for users to independently develop network applications and program network devices. It is noticed that such solutions are dominated by current mainstream network equipment vendors and are the most widely used.

1.3.2   Network Function Virtualization

Current network services rely on proprietary appliances and different network devices that are diverse and purpose-built [98, 115]. This situation induces the so-called network ossification problem, which prevents the operation of service additions and network upgrades. To address this issue and reduce capital expenditures (CapEx) and operating expenditures (OpEx), virtualization has emerged as an approach to decouple the software networking processing and applications from their supported hardware and allow network services to be implemented as software [14, 93]. Leveraging virtualization technologies, ETSI Industry Specification Group proposed Network Functions Virtualization (NFV) to virtualize the network functions that are previously carried out by some proprietary dedicated hardware [13, 125]. By decoupling the network functions from the underlying hardware appliances, NFV provides flexible provisioning of software-based network functionalities on top of an optimally shared physical infrastructure. It addresses the problems of operational costs of managing and controlling these closed and proprietary appliances by leveraging low cost commodity servers.

At the same time, the next-generation mobile, enterprise, and IoT networks are introducing the concept of computing capabilities being pushed at the network edge, in close proximity of the users. However, the heavy footprint of today's NFV platforms prevents them from operating at the network edge. For this reason, a new, lightweight virtualization technology container has been proposed. It could typically avoid the hardware requirements and overheads associated with hypervisors and VMs. Containers incur significantly lower overhead than traditional VMs and can be deployed in any Linux environment (available from commodity routers to high-end servers) with similar performances to the host machine. Similar to specialized VMs, containers also allow much higher network function-to-host density and smaller footprint at the cost of reduced isolation. Using containers, commodity compute devices (or public cloud VMs) are able to host up to hundreds of virtual network functions (vNFs) as shown in [18, 32].

In addition, NFV and SDN are closely related and highly complementary to each other. NFV can serve SDN by virtualizing the SDN controller (which can be regarded as a network function) to run on cloud, thus allowing dynamic migration of the controllers to the optimal locations. So combining SDN with NFV is a good way to improve both NFV's and SDN's ability. SDS could be implemented to contain a control module, forwarding devices, and NFV platform at the edge of the network. The logic of packet forwarding is determined by the SDN controller and is deployed in the forwarding devices through forwarding tables. Efficient protocols, e.g., Openflow [9, 65], can be utilized as standardized interfaces in communicating between the centralized controller and distributed forwarding devices. The NFV platform leverages commodity servers to implement high bandwidth NFs at low cost. Hypervisors run on the servers to support the VMs that implement the NFs. Accordingly, SDS will be able to achieve customizable and programmable data plane processing functions such as middle-box of recalls, IDSes, proxies, which are running as software within virtual machines, where NFs are delivered to the network operator as pieces of pure software. In turn, SDN serves NFV by providing programmable network connectivity between VNFs to achieve optimized traffic engineering and steering [23, 40].

1.3.3 Machine Learning

With the prosperous development of the Internet, networking research has attracted a lot of attention in the past several decades both in academia and industry.

Researchers and network operators can face various types of networks (e.g., wired or wireless) and applications (e.g., network security and live streaming [104]). Each network application also has its own features and performance requirements, which may change dynamically with time and space. Because of the diversity and complexity of networks, specific algorithms are often built for different network scenarios based on the network characteristics and user demands. Developing efficient algorithms and systems to deal with complex problems in different network scenarios is a challenging task.

Recently, machine learning (ML) techniques have made breakthroughs in a variety of application areas, such as bioinformatics, speech recognition, and computer vision. Machine learning tries to construct algorithms and models that can learn to make decisions directly from data without following predefined rules. Dealing with complex problems is one of the most important advantages of machine learning. Since the network field often sees complex problems that demand efficient solutions, it is promising to bring machine learning algorithms into the network domain to leverage the powerful ML abilities for higher network performance.

The ML technology especially reinforcement learning (RL) is an important method to solve the network challenges. Reinforcement learning is different from supervised learning and unsupervised learning, the kind of learning studied in most current research in the field of machine learning [87]. Reinforcement learning (RL) is learning by interacting with an environment. An RL agent observes the state of the environment, performs actions in the environment, and receives reward feedback from the environment to autonomously understand the environment and complete tasks. The agent learns from the consequences of its actions to obtain the optimal strategy, so that the cumulative reward is the largest. The classic study of reinforcement learning is based on the Markov decision process (MDP)[85].

In the field of wireless sensor networks (WSN), many of the WSN applications rely heavily on fast, efficient, and reliable data communications. For example, the authors of [81] introduce a near-optimal reinforcement learning framework for energy-aware sensor communications. The problem is formulated as average throughput maximization per total consumed energy in sensor communications. A near-optimal transmission strategy was obtained based on a reinforcement learning framework. This strategy chooses the optimal modulation level and transmission power while adapts to the incoming traffic rate, buffer, and channel conditions.

In addition, many have used online learning techniques in automated building control systems in WSN, though the solutions tend to require significant computation and consequently centralized support. In order to avoid the need for centralization, this system must be able to learn in a distributed manner. In this case, it is possible to apply multi-agent reinforcement learning to such environment [92].

1.3.4 Cognitive Radio

With the rapid deployment of new wireless devices and applications, the last decade has witnessed a growing demand for wireless radio spectrum. However, the fixed spectrum assignment policy becomes a bottleneck for more efficient spectrum utilization, under which a great portion of the licensed spectrum is severely under-utilized. The inefficient usage of the limited spectrum resources urges the spectrum regulatory bodies to review their policy and start to seek for innovative communication technology that can exploit the wireless spectrum in a more intelligent and flexible way. The concept of cognitive radio(CR) is proposed to address the issue of spectrum efficiency and has been receiving an increasing attention in recent years, since it equips wireless users the capability to optimally adapt their operating parameters according to the interactions with the surrounding radio environment. There have been many significant developments in the past few years on cognitive radios.

A cognitive radio (CR) is a software defined radio (SDR) that additionally senses its environment, tracks changes, and reacts upon its findings. It differs from traditional communication paradigms in that the radios/devices can adapt their operating parameters, such as transmission power, frequency, modulation type, etc., to the variations of the surrounding radio environment [55]. Before CRs adjust their operating mode to environment variations, they must first gain necessary information from the radio environment. This kind of characteristics is referred to as cognitive capability [41], which enables CR devices to be aware of the transmitted waveform, radio frequency (RF) spectrum, communication network type/protocol, geographical information, locally available resources and services, user needs, security policy, and so on.

A typical duty cycle of CR includes detecting spectrum white space, selecting the best frequency bands, coordinating spectrum access with other users, and vacating the frequency when a primary user appears. Such a cognitive cycle is supported by three functions: spectrum sensing and analysis, spectrum management and handoff, and spectrum allocation and sharing.

Although CR can provide military with adaptive, seamless, and secure communications, the transmissions of secondary users are suspended during a spectrum handoff. They will experience longer packet delay. A good way to alleviate the performance degradation due to long delay is to reserve a certain number of channels for potential spectrum handoff [130]. In additions, considering the frequency agility and adaptive bandwidth, the concept of time-spectrum block is introduced in [124], and a distributed protocol is developed to solve the spectrum allocation problem which enables each node to dynamically choose the best time-spectrum block based only on local information.

1.3.5 Cloud and Edge Computing

The last decade has seen cloud computing emerging as a new paradigm of computing. Its vision is the centralization of computing, storage and network management in the Clouds, referring to data centers, backbone IP networks, and cellular core networks [29, 128]. The vast resources available in the Clouds can then be leveraged to deliver elastic computing power and storage to support resource-constrained end-user devices. Cloud computing has been driving the rapid growth of many Internet companies. For example, the cloud business has risen to be the most profitable sector for Amazon [118], and Dropbox's success depended highly on the cloud service of Amazon.

However, in recent years, a new trend in computing is happening with the function of clouds being increasingly moving towards the network edges [12]. It is estimated that tens of billions of Edge devices will be deployed in the near future, and their processor speeds are growing exponentially, following Moore's Law. Harvesting the vast amount of the idle computation power and storage space distributed at the network edges can yield sufficient capacities for performing computation intensive and latency-critical tasks at mobile devices. This paradigm is called mobile edge computing (MEC) [82].

The common denominator in these edge paradigms is the deployment of cloud computing-like capabilities at the edge of the network. Edge data centers, which are owned and deployed by infrastructure providers, implement a multi-tenant virtualization infrastructure. Any customer—from third-party service providers to end users and the infrastructure providers themselves—can make use of these data centers' services. In addition, while edge data centers can act autonomously and cooperate with one another, they are not disconnected from the traditional cloud. It is therefore possible to create a hierarchical multi-tiered architecture, interconnected by a network infrastructure. Besides, we have to consider the potential existence of an underlying infrastructure, or core infrastructure, that provide various support mechanisms, such as management platforms and user registration services. Finally, one trust domain (i.e., edge infrastructure that is owned by a infrastructure provider) can cooperate with other trust domains, creating an open ecosystem where multitude of customers can be served.

The benefits of deploying cloud services at the edge of mobile networks especially in 5G include low latency, high bandwidth, and access to radio network information and location awareness. Thanks to this, it will be possible to optimize existing mobile infrastructure services, or even implement novel ones. An example is the mobile edge scheduler [24], which minimizes the mean delay of general traffic flows in the LTE downlink. Besides, the MEC ISG has suggested that this virtualization infrastructure should host not only MEC services, but also other related services such as network function virtualization (NFV) and software defined networking (SDN) [46].

1.3.6 Microservices

When one is developing a server-side enterprise applications, they must support a variety of different clients including desktop browsers, mobile browsers, and native mobile applications. The application might also expose an API for 3rd parties to consume. It might also integrate with other applications via either web services or a message broker. For example, the application handles requests (HTTP requests and messages) by executing business logic; accessing a database; exchanging messages with other systems; and returning a HTML/JSON/XML response. There are logical components corresponding to different functional areas of the application. However, the following forces make it very challenging to design the deployment architecture of applications [4, 74]:

- There is a team of developers working on the application;
- New team members must quickly become productive;
- The application must be easy to understand and modify;
- Someone wants to practice continuous deployment of the application;
- Someone must run multiple instances of the application on multiple machines in order to satisfy scalability and availability requirements;
- Someone wants to take advantage of emerging technologies (frameworks, programming languages, etc).

To this end, successful microservices deployments rely on an architecture that structures the application as a set of loosely coupled, collaborating services, for the following demands [74]:

- Highly maintainable and testable—enables rapid and frequent development and deployment;
- Loosely coupled with other services—enables a team to work independently the majority of time on their service(s) without being impacted by changes to other services and without affecting other services;
- Independently deployable—enables a team to deploy their service without having to coordinate with other teams;
- Capable of being developed by a small team—essential for high productivity by avoiding the high communication head of large teams.

Services communicate using either synchronous protocols such as HTTP/REST or asynchronous protocols. Services can be developed and deployed independently of one another. Each service has its own database in order to be decoupled from other services. For example, the fictitious e-commerce application, which belongs to an e-commerce application that takes orders from customers, verifies inventory and available credit, and ships them. Such an application consists of several components including the StoreFrontUI to implement the user interface, along with some backend services for checking credit, maintaining inventory and shipping orders, and can be easily decomposed into multiple microservices.

Computer microservices can be implemented in different programming languages and might use different infrastructures. Therefore the most important technology choices are the way microservices communicate with each other (synchronous, asynchronous, UI integration) and the protocols used for the communication. In a traditional system most technology choices like the programming language impact the whole systems. Therefore the approach for choosing technologies is quite different. When each service instance is paired with an instance of a reverse proxy server, called a service proxy, sidecar proxy, or sidecar, they build a service mesh. In a service mesh, the service instance and sidecar proxy share a container, and the containers are managed by a container orchestration tool such as Kubernetes. The service proxies are responsible for communication with other service instances and can support capabilities such as service (instance) discovery, load balancing, authentication and authorization, secure communications, and others. However, the architecture introduces additional complexity and new problems to deal with, such as network latency, message formats, load balancing, and fault tolerance. All of these problems have to be addressed at scale [83].

As a whole, microservices are a software development technique. A variant of the service-oriented architecture (SOA) architectural style that structures an application as a collection of loosely coupled services. In a microservices architecture, services are fine-grained and the protocols are lightweight. The benefit of decomposing an application into different smaller services is that it improves modularity. This makes the application easier to understand, develop, test, and become more resilient to architecture erosion. It parallelizes development by enabling small autonomous teams to develop, deploy, and scale their respective services independently. It also allows the architecture of an individual service to emerge through continuous refactoring. Microservice-based architectures enable continuous delivery and deployment [4].

1.4 Aim of this Brief

SDS uses software to automate and virtualize the main components of your IT architecture: compute, networking, and storage. There are a number of benefits of software defined technology and the aim of this brief is to highlight five major benefits of SDS. We call them the 5As as the followings [57]:

Accuracy   In transitioning to a software driven enterprise, the network framework will become programmable and automated, thus eliminating human (or even machine) error. Think about the implications of this: no longer will the traffic traversing your network be dependent upon the hardware you are running. Rather, the traffic will be routed intelligently, via software that is smarter than a switch or router. The application, once the gravy on top of the meal, now becomes the entire meal itself. It knows what it needs to do, and it takes the most efficient path to get there, allowing for intelligent accuracy that reduces the need for constant babysitting of your network by IT staff, as well as cuts your costs and improves your operating efficiency.

Adaptability No more reliance on hardware vendors who have you locked in, with big capital expenditures for upgrades that you cannot afford. Your network or data storage will now be software-enabled or virtualized so that you can easily switch from environment to environment without the hassle of a "rip and replace" initiative or hiring extra staff to manage an environment change. This allows your company to scale on a very low margin, saving you time and money (and of course, headaches.)

Agility Agility is the ease in which your organization's data computation can navigate complex environments quickly, and according to your enterprise's specific needs. According to an article on [111], "Business agility is the new currency for valuing technology in the enterprise and increased agility is what virtualization delivered and compute clouds promise."

With software defined systems, you can move between one system and another, switching environments whenever you choose. You can set up your applications instantly over a software defined system, and you can take them down instantly as well. This agile environment optimizes the user experience while greatly reducing costs.

Alignment All of your resources within your infrastructure will be completely aligned, rather than disparate hardware and software that requires intensive (and costly) IT maintenance. In fact, the alignment benefit from software defined technology will be a driving factor in IT innovation.

Rather than incurring a huge expense for IT staff dedicated to support a mismatch of vendors, equipment, and software, IT will be looked upon to innovate—driving profitability with a more strategic approach. The role of IT will be elevated to oversee infrastructure that is driven by business policies and priorities, rather than being consumed with fixing problems and putting out fires. Software defined systems enable IT to be proactive, rather than reactive.

Assurance Every organization has policy and compliance requirements that must be in place and monitored. With software defined systems in place, your organization will have a much higher degree of confidence that your entire infrastructure is compliant with the standards and regulations that your organization must adhere to.

As we can see, these 5As guide us toward a congruent and optimal infrastructure that promises to propel network systems into the next generation of information system with ease. In what follows, this book introduces novel solutions that focus on the sensing, communications, and networking in the ICT with SDS. We examine the latest research advances in SDS for next-generation wired, wireless, and sensor networks. We present efficient algorithms, protocols, and network architectures to improve efficiency and practicability of SDS, and comprehensively illustrate the various aspects of modeling, analysis, design, management, deployment, and optimization of algorithms, protocols, and architectures of green communications and networking.

Chapter 2
Software Defined Sensing

Abstract After a decade of extensive research on application-specific WSNs, the recent development of information and communication technologies makes it practical to realize SDSNs, which are able to adapt to various application requirements and to fully explore the resources of WSNs. A sensor node in SDSN is able to conduct multiple tasks with different sensing targets simultaneously. A given sensing task usually involves multiple sensors to achieve a certain quality-of-sensing, e.g., coverage ratio. It is significant to design an energy-efficient sensor scheduling and management strategy with guaranteed quality-of-sensing for all tasks. To this end, three issues shall be considered: (1) the subset of sensor nodes that shall be activated, i.e., sensor activation, (2) the task that each sensor node shall be assigned, i.e., task mapping, and (3) the sampling rate on a sensor for a target, i.e., sensing scheduling. In this chapter, they are jointly considered and formulated as a mixed-integer with quadratic constraints programming (MIQP) problem, which is then reformulated into a mixed-integer linear programming (MILP) formulation with low computation complexity via linearization. To deal with dynamic events such as sensor node participation and departure, during SDSN operations, an efficient online algorithm using local optimization is developed. Simulation results show that the proposed online algorithm approaches the globally optimized network energy efficiency with much lower rescheduling time and control overhead.

2.1   Programmable Sensors

Wireless Sensor Networks (WSNs) have been widely deployed for a wide span of applications including surveillance, tracking, and controlling. While many efforts have been made to enhance the applicability and performance of WSNs from different layers, they mainly consider the case that a WSN is dedicated for one sensing task. Such an application-specific WSN is prone to

© The Author(s), under exclusive license to Springer Nature Switzerland AG 2020   17
D. Zeng et al., *Software Defined Systems*, SpringerBriefs in Computer Science,
https://doi.org/10.1007/978-3-030-32942-6_2

1. high deployment cost: multiple WSNs for respective tasks may be deployed in the same area,
2. low service reutilization: different vendors develop their WSNs in an isolated manner without sharing common functionalities, and
3. difficult hardware recycling: altering existing code on single-task sensor nodes is difficult, highly error-prone, and costly [63].

A software defined (or programmable) sensor node equipped by several different types of sensors is able to conduct different sensing tasks according to the programs deployed and activated. Such kind of sensor node prototypes have been practically realized. Miyazaki et al. [70, 71] implemented the software defined sensor nodes that can dynamically change their functions upon specific sensing task requirements at runtime.

2.2 Software Defined Sensor Networks

By exploring the programmability of software defined sensor nodes, software defined sensor networks (SDSNs) emerge as a compelling solution to address the limitations of application-specific WSNs [63]. An SDSN consists of a number of programmable sensor nodes whose functionalities can be dynamically configured by injecting different programs.

Figure 2.1 shows a general architecture of SDSN, which consists of a sensor control server and a set of software defined sensor nodes. Any node is integrated with multiple sensors of various types, e.g., ultrasonic sensor, photoelectric sensor, infrared sensor, etc., each of which is responsible for a specific sensing task for a corresponding group of targets in its sensing area. The targets are classified by the

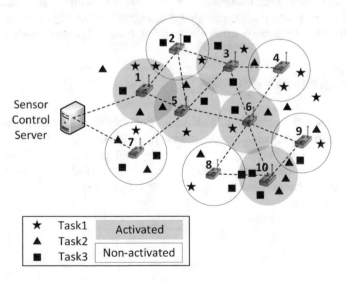

Fig. 2.1 An example of software defined sensor network with three types of sensing targets

sensing tasks as depicted by different notations, i.e., ★, ▲, and ■ in Fig. 2.1. At each time instance, a sensor node can exactly sense one target. In software defined sensor node, operation system is required for the management of the sensor resources [71]. Modern sensor node OS like TinyOS [59] usually supports multi-task that can execute independently and non-preemptively. Multiple tasks with different sensing targets may be issued to the SDSN. For each task, there is a related program. Only the sensors that have been loaded and activated with the corresponding program are able to sense the related targets within its coverage. For example, a sensor node can conduct vibration and heat detection using photoelectric and infrared sensors, respectively, provided that both corresponding programs are loaded.

Energy efficiency is always a critical issue in WSNs with battery-powered nodes and SDSNs are without exception. Intuitively, the less sensors are activated, the less energy shall be consumed. However, a sensing task requires certain level of quality-of-sensing, e.g., coverage ratio, which is a commonly adopted quality-of-sensing metric describing the portion of targets covered by the reprogrammed sensors [112]. Ideally, for multi-task SDSNs, the computation and storage capacity limitations of sensors and the requirements of sensing tasks shall be jointly taken into account. Therefore, in this chapter, we study a minimum-energy sensor activation problem in multi-task SDSNs with guaranteed quality-of-sensing. We first derive the effective sensing rate that can be achieved by collaborative sensing in closed-form, based on which we formulate the minimum-energy sensor activation problem as a mixed-integer with quadratic constraints programming (MIQP) problem by jointly considering sensor activation and task mapping. Based the formulation, we further propose an efficient online algorithm to deal with dynamic events during runtime of SDSNs.

2.3 Task Scheduling in Software Defined Sensor Networks

We consider an SDSN like the one shown in Fig. 2.1, where each sensor node is equipped with multiple sensors with different sensing capabilities, e.g., temperature, humidity, light, vibration, etc. Such an SDSN can run multiple tasks, each of which is explicitly specified with a set of sensing targets and the quality-of-sensing requirement, e.g., minimum *coverage ratio*. In this paper, we are mainly interested in point coverage, where a target is a detectable spatial point-of-interest (PoI) with location information [47, 112, 113]. For example, in a structural health monitoring (SHM) task, a target refers to a vulnerable point that may be critical to the health of a monitored structure, e.g., a skyscraper or a bridge, as discussed in [7]. Let S and T denote the sensor set and task set, respectively. For a task $t \in T$, its sensing target set is represented by G_t and the required coverage ratio is denoted by δ_t.

Without loss of generality, we assume that both sensor nodes and the sensing targets are randomly distributed in the network area. Target $g \in G_t$ of a task $t \in T$

can be monitored by sensor $s \in S$ only if residing within the sensing range of s. This fact can be expressed by notations

$$v_{stg} = \begin{cases} 1, \text{ if target } g \text{ of task } t \text{ is within the sensing} \\ \quad \text{range of sensor node } s, \\ 0, \text{ otherwise.} \end{cases}$$

Note that, a sensor node may have different sensing ranges for different tasks. Once the locations of sensor nodes and the sensing targets are known, the values of v_{stg} will be determined.

In SDSNs, only the sensor nodes $s \in S$ loaded with a program for task $t \in T$ can sense t's targets. Therefore, to conduct a sensing task on a sensor node first requires that the corresponding program is stored on the sensor node. Different programs are with different program sizes and hence are with diverse storage requirements. We consider a homogeneous sensor network that all the sensors are with the same storage capacity, which is normalized as 1.0 throughout this paper. Consequently, let $z_t, 0 < z_t \leq 1, t \in T$ denote the normalized program size of task t.

To ensure the sensing accuracy to a target, we must guarantee the minimum sensing rate requirement, which refers to how often a sensing operation shall be conducted to a target. The sensing rate is usually specified in the task requirement. Only the minimum sensing rate to a target is satisfied, we can say that the target is covered. Besides, each sensing operation also needs certain sensing and computation duration. The need for non-negligible sensing durations to obtain useful information is due to noises in the measurement process and the probabilistic nature of the phenomena under observation [122]. For a task $t \in T$, let us use f_t and c_t to denote the required minimum sensing rate and duration to its target, respectively. For example, to sense the vibration at a PoI (i.e., target) of a SHM task t, we need sensing rate and duration as 300 Hz and 1 ms, respectively, i.e., $f_t = 300$ Hz and $c_t = 1$ ms. We treat a target g of sensing task t as "covered" provided that "the effective sensing rate," denoted as f_{tg}^e, from multiple sensors exceeds the minimum sensing rate requirement f_t. Any involved sensor s independently makes the corresponding sensing activity following a Poisson process with a rate f_{stg}.

Multiple tasks need to be handled by an SDSN at the same time. A subset of the sensor nodes in the network shall be activated for these tasks. It has been widely proved that careful scheduling on the sensor activation is a promising way to conserve the power consumption [11, 51, 117, 123]. Therefore, a sensor could be in activated mode or sleeping mode, with power consumption P_a and P_s, respectively. In an SDSN, a sensor node may be loaded with multiple programs for different tasks. Therefore, we are interested in which sensors shall be activated and which tasks shall be assigned to each of them to minimize the sensing power consumption while guaranteeing the quality-of-sensing for each task.

2.4 SDSN Task Scheduling Problem Formulation

2.4.1 Coverage Ratio Constraints

We first consider the quality-of-sensing requirement for a sensing task in SDSNs. A sensor $s \in S$ can sense the target $g \in G_t$ of task $t \in T$ if and only if (1) target t is within the sensing range of s, i.e., $v_{stg} = 1$ and (2) sensing task t is scheduled on s. We define a binary variable α_{st} to denote whether sensor s is scheduled with task t or not as follows:

$$\alpha_{st} = \begin{cases} 1, & \text{if sensor } s \text{ is scheduled with task } t, \\ 0, & \text{otherwise.} \end{cases}$$

By letting β_{stg} denote whether sensor s is able to sense target g of task t, we have

$$\beta_{stg} = \alpha_{st} v_{stg}, \forall s \in S, t \in T, g \in G_t. \tag{2.1}$$

For any sensing target counted as covered (i.e., $\gamma_g = 1, \forall t \in T, g \in G_t$), the minimum sensing rate to it must be reserved to ensure the sensing accuracy. The corresponding program shall be invoked periodically according to the task sensing rate requirement. Note that the sensing rate f_{stg} for a target is first determined by β_{stg} as:

$$0 \le f_{stg} \le \beta_{stg} \cdot f_t, \forall s \in S, t \in T, g \in G_t, \tag{2.2}$$

which indicates that only when sensor s is able to sense target g of task t, f_{stg} can be allocated with a value larger than 0.

Multiple sensors may be able to sense one target cooperatively. For example, as shown in Fig. 2.1, the target of *Task 2* in black triangle can be sensed by both sensor 1 and sensor 5 in their overlapped coverage. Note that the effective sensing rate is not simply obtained by summing up the sensing rates contributed by the collaborative sensors, because any sensing event that meets an on-going one for the same target will be considered duplicate and should be ignored. The derivation of the effective sensing rate is given in the theorem below.

Theorem 2.1 *The effective sensing rate in collaborative sensing for target g of sensing task t is*

$$f_{tg}^e = \frac{\sum_{s \in S} f_{stg}}{1 + c_t \sum_{s \in S} f_{stg}}. \tag{2.3}$$

Proof Recall that the sampling events are triggered as a Poisson process by each sensor independently. The overall sensing activities can thus be considered as a compound Poisson process with rate $f = \sum_{s \in S} f_{stg}$. When a sensing activity of task t (with a fixed duration c_t) is on-going, any newly triggered event in this duration will be considered as duplicated. Therefore, the collaborative sensing for target g of task t can be modeled as an M/D/1/1 queue. According to the transient analysis by Garcia et al. on M/D/1/N queue [30], we can derive the stationary probability distribution function for our M/D/1/1 queue as:

$$\pi_0 = \frac{1}{1+\rho},$$

$$\pi_1 = 1 - \frac{1}{1+\rho},$$

(2.4)

where

$$\rho = c_t \sum_{s \in S} f_{stg}.$$

(2.5)

A sensing event is considered as effective only when the "queuing system" is in state π_0. Therefore, the effective sensing rate f_{tg}^e can be calculated as:

$$f_{tg}^e = f\pi_0 = \frac{\sum_{s \in S} f_{stg}}{1 + c_t \sum_{s \in S} f_{stg}}.$$

(2.6)

∎

Based on the result from Theorem 1, we define a binary variable γ_{tg} to represent if the effective sensing rate f_{tg}^e exceeds the required sensing rate f_t, i.e.,

$$\gamma_{tg} = \begin{cases} 1, \text{ if } \dfrac{\sum_{s \in S} f_{stg}}{1 + c_t \sum_{s \in S} f_{stg}} \geq f_t, & \forall t \in T, g \in G_t. \\ 0, \text{ otherwise.} \end{cases}$$

(2.7)

We can rewrite (2.7) by the following two inequalities:

$$\gamma_{tg} \cdot \left(\sum_{s \in S} f_{stg} - f_t - c_t f_t \sum_{s \in S} f_{stg} \right) \geq 0, \forall t \in T, g \in G_t$$

(2.8)

and

$$(1 - \gamma_{tg}) \cdot \left(f_t + c_t f_t \sum_{s \in S} f_{stg} - \sum_{s \in S} f_{stg} \right) \geq 0, \forall t \in T, g \in G_t$$

(2.9)

It can be verified that if $\sum_{s \in S} f_{stg} - f_t - c_t f_t \sum_{s \in S} f_{stg} \geq 0$, i.e., $\dfrac{\sum_{s \in S} f_{stg}}{1 + c_t \sum_{s \in S} f_{stg}} \geq f_t$, then γ_{tg} must be 1 in order to satisfy (2.9). Similarly, if $\sum_{s \in S} f_{stg} - f_t -$

$c_t f_t \sum_{s \in S} f_{stg} \leq 0$, i.e., $\frac{\sum_{s \in S} f_{stg}}{1 + c_t \sum_{s \in S} f_{stg}} \leq f_t$, then γ_{tg} must be 0 in order to satisfy (2.8).

To ensure the quality-of-sensing for each sensing task, e.g., $t \in T$, a coverage ratio δ_t shall be achieved. This leads to the following constraints:

$$\sum_{g \in G_t} \gamma_{tg} / |G_t| \geq \delta_t, \forall t \in T. \tag{2.10}$$

2.4.2 Schedulability Constraints

A sensor node may have multiple targets to sense. A sensing target $g \in G_t, \forall t \in T$ requires certain sensing rate f_{stg} and duration c_t on a sensor node $s \in S$.

According to [96], the following constraints must be satisfied for all the sensor nodes:

$$\sum_{t \in T} \sum_{g \in G_t} c_t \cdot f_{stg} \leq 1, \forall s \in S, \tag{2.11}$$

to ensure the multi-task schedulability.

2.4.3 Storage Constraints

Due to the sensor node storage capacity limitation, the total storage requirement for all the tasks mapped onto a sensor node shall not exceed its storage capacity. Then, we have

$$\sum_{t \in T} \alpha_{st} \cdot z_t \leq 1, \forall s \in S. \tag{2.12}$$

2.4.4 Problem Formulation

If a sensor node is scheduled for at least one task, it must be activated to conduct the sensing operations. Let binary $a_s, s \in S$ denote whether a sensor node is activated or not, i.e.,

$$a_s = \begin{cases} 1, & \text{if sensor } s \text{ is activated,} \\ 0, & \text{otherwise.} \end{cases}$$

Then, we have:

$$\sum_{t \in T} \alpha_{st}/|T| \le a_t \le \sum_{t \in T} \alpha_{st}, \forall s \in S. \qquad (2.13)$$

Obviously, from (2.13), we can see that $a_s \equiv 1$ if $\exists t \in T, \alpha_{st} = 1$, and $a_s \equiv 0$ only if $\forall t \in T, \alpha_{st} = 0$.

Our objective to minimize the total sensing power consumption is equivalent to minimizing the number of sensors that shall be activated, i.e., $\sum_{i \in I} a_i$. By summarizing all the above issues together, we may obtain an **MIQP** formulation as: Our objective of minimizing the total sensing power consumption can be represented as $\sum_{s \in S} a_s P_a + (1 - a_s) P_s$. By summarizing all the above constraints together, we obtain an **MIQP** formulation as:

MIQP:

$$\min : \sum_{s \in S} a_s P_a + (1 - a_s) P_s,$$

$$\text{s.t. :}\quad (2.1), (2.2), (2.8)\text{--}(2.13).$$

2.5   Linearization

It has been proved that MIQP is NP-Hard to solve [8]. Fortunately, we observe that the constraints (2.8) and (2.9) are non-linear because of the products of variables. To linearize these constraints and lower the computation complexity, we define a new variable u_{stg} as follows:

$$u_{stg} = f_{stg} \gamma_{tg}, \forall s \in S, t \in T, g \in G_t, \qquad (2.14)$$

which can be equivalently replaced by the following linear constraints:

$$0 \le u_{stg} \le f_{stg}, \forall s \in S, t \in T, g \in G_t, \qquad (2.15)$$

$$f_{stg} + \gamma_{tg} - 1 \le u_{stg} \le \gamma_{tg} f_j, \forall s \in S, t \in T, g \in G_t. \qquad (2.16)$$

The constraints (2.8) and (2.9) then can be written in a linear form as:

$$\sum_{s \in S} u_{stg} - \gamma_{tg} f_t - c_t f_t \sum_{s \in S} u_{stg} \ge 0, \forall t \in T, g \in G_t \qquad (2.17)$$

and

$$f_t + c_t f_t \sum_{s \in S} f_{stg} - \sum_{s \in S} f_{stg} - \gamma_{tg} f_t - c_t f_t \sum_{s \in S} u_{stg}$$

$$+ \sum_{s \in S} u_{stg} \geq 0, \forall t \in T, g \in G_t, \tag{2.18}$$

respectively.

Now, we can linearize the **MIQP** problem into a mixed-integer linear programming (**MILP**) as:

MILP:

$$\min : \sum_{s \in S} a_s P_a + (1 - a_s) P_s,$$

$$\text{s.t.} : (2.1), (2.2), (2.10)–(2.13), (2.15)–(2.18).$$

2.6   Online Management in Software Defined Sensor Networks

In SDSNs, there are two kinds of network dynamics, referring to applications and sensor nodes, respectively. Applications may arrive during network operation and existing sensor nodes may depart because of power depletion or other events, and new nodes may be deployed. In this section, we present online algorithm to deal with these dynamic events.

2.6.1   Application Dynamics

As an SDSN may be released to tenants, who may submit new application or application requirement to sensor control server periodically. When sensor control server receives such a request, different from initial deployment, it shall consider the resource availability in the network as some nodes have already been deployed with certain applications. Fortunately, the sensor control server has global information of the whole network, with which it is able to apply global optimization similar to the initial deployment as discussed above.

When a new task t' comes, the control server updates the task set as $T = T \cup \{t'\}$ and make new task assignment decisions $a_{st}, \forall s \in S, t \in T$ for the new task set. The control server shall still consider coverage ratio constraints, schedulability constraints, and storage constraints as discussed in Sect. 2.4. Besides, to avoid task migration on the sensor nodes that have been deployed with tasks, we further incorporate the following task-reserving constraints as:

$$a_{st} \geq a_{st}^{\text{old}}, \forall s \in S, t \in T, \tag{2.19}$$

where a_{st}^{old} denotes the task assignment decision before the arrival of new application t'. Equation (2.19) indicates that if task t has been assigned to node s, i.e., $a_{st}^{\text{old}} = 1$, such assignment shall still be reserved after resource allocation for the new task. Note that we do not enforce reserving on the sensing frequency and therefore it is possible to tune the sensing frequencies of the assigned tasks so as to accommodate the new task, if necessary. Summing up all the constraints, we get the MILP for new application as:

MILP-NEW-APP:

$$\min : \sum_{s \in S} a_s P_a + (1 - a_s) P_s,$$

$$\text{s.t.} : (2.1), (2.2), (2.10)\text{--}(2.13), (2.15)\text{--}(2.19).$$

2.6.2 Sensor Node Dynamics

On the other hand, in a practical wireless sensor network, sensor nodes are powered by batteries with limited capacity, and usually they cannot be recharged. Therefore, new nodes will be deployed to compensate the portion of sensors that have exhausted their batteries. In this section, we consider a dynamic network where existing nodes will depart because of power depletion, and new nodes will be deployed periodically. An intuitive method to deal with network dynamic is to apply the global optimization proposed in last section each time when joining or leaving events happen. Although such a method always leads to the minimum number of active sensors, it suffers from three weaknesses. First, frequent execution of global algorithm incurs high computational complexity at the sink. Second, the sink needs to collect the information of joining and leaving nodes for global optimization, and then deploys the results to all nodes, which would generate too many control messages that also consume a large amount of power. Finally, the control messages between sink and sensors are exchanged over a multi-hop network, and the resulting delay cannot be ignored, especially in a large wireless sensor network where packets may travel through hundreds of hops from sink to the farthest node.

Instead of adopting global optimization to deal with network dynamic, we propose an online algorithm using local optimization with low complexity.

2.6.2.1 Participation

We first consider the case that a sensor node is deployed as a participator to the network. Although letting this node stay inactive will not degrade the quality-of-

Fig. 2.2 An example of
participation

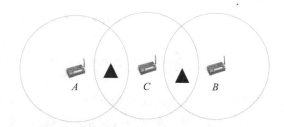

sensing, many opportunities of reducing the number of active nodes are missed. As an example shown in Fig. 2.2, two targets are sensed by two active sensor nodes A and B, respectively. When a new node C is deployed and its sensing range is large enough to cover two targets, we can deactivate nodes A and B to keep only one active sensor. To exploit such opportunities for performance improvement when new nodes are deployed, we propose an online algorithm to deal with participation events.

Step 1: When a new node x is deployed, it first registers its information including the location, resource capacity, sensing units to the sensor control server.

Step 2: With the registered information, the sensor control server can discover the potential sensing targets within its sensing range. Centered at these potential targets, the sensor control server can also derive a set of sensor nodes that can reach these targets, e.g., sensor nodes A and B in Fig. 2.2.

Step 3: Based on the above information, the sensor control server can establish a subgraph including a set S' of sensors, and a set G'_t of targets for each task $t \in T$, as well as the corresponding quality-of-sensing δ'_t that should be satisfied in the subgraph, i.e.,

$$\delta'_t \equiv \sum_{g \in G'_t} \bar{\gamma}_{tg}/|G'_t|, \tag{2.20}$$

where $\bar{(\cdot)}$ denotes existing value of any variable $(\cdot)$.

Step 4: The sensor control server then solve a local optimization on the subgraph to reduce the number of active sensor nodes while guaranteeing the quality-of-sensing of each task. The formulation of local optimization **Local_OPT** is shown as follows:

$$\min : \sum_{s \in S'} a_s P_a + (1 - a_s) P_s$$

$$\text{s.t.} : 0 \leq f_{stg} \leq \alpha_{st} \cdot \upsilon_{stg} \cdot f_t, \forall s \in S', t \in T, g \in G'_t,$$

$$\sum_{g \in G'_t} \gamma_{tg}/|G'_t| \geq \delta'_t, \forall t \in T,$$

$$\sum_{t \in T} \sum_{g \in G'_j} c_t \cdot f_{stg} \leq 1, \forall s \in S',$$

$$\sum_{t \in T} \alpha_{st} \cdot z_t \leq 1, \forall s \in S',$$

$$\sum_{t \in T} \alpha_{st}/|T| \leq a_s \leq \sum_{t \in T} \alpha_{st}, \forall s \in S',$$

$$0 \leq u_{stg} \leq f_{stg}, \forall s \in S', t \in T, g \in G'_t,$$

$$f_{stg} + \gamma_{tg} - 1 \leq u_{stg} \leq \gamma_{tg} f_j, \forall s \in S', t \in T, g \in G'_t,$$

$$\sum_{s \in S} u_{stg} - \gamma_{tg} f_t - c_t f_t \sum_{s \in S} u_{stg} \geq 0, \forall t \in T, g \in G'_t,$$

$$f_t + c_t f_t \sum_{s \in S} f_{stg} - \sum_{s \in S} f_{stg}$$

$$- \gamma_{tg} f_t - c_t f_t \sum_{s \in S} u_{stg} + \sum_{s \in S} u_{stg} \geq 0, \forall t \in T, g \in G_t.$$

Since the number of variables and constraints in local optimization is very limited, it can be easily solved even by resource-constrained sensor nodes.

Step 5: After solving the local optimization, if node x needs to become active such that other two or more sensors can be deactivated, corresponding commands will be disseminated to the involved sensors to adjust their sensing strategy. Otherwise, node x remains inactive, and sensors in set S' still use the existing sensing strategy.

2.6.2.2   Departure

We then consider that case that a sensor node departs from the network because of power depletion. We also adopt a local optimization to compensate the sensing rate of the departed node with the minimize number of sensors. However, if the local sensor nodes are not able to guarantee the quality-of-sensing, we resort to the global optimization at the sink. The corresponding process is elaborated as follows:

Step 1: When a node, say x, assigned with sensing tasks departs from the network. Its neighbor shall be able to discover such event and reports it to the sensor control server. For sensing data reporting, the neighborhood information between activated sensor nodes is already maintained in wireless sensor networks and therefore it is easy to detect the departure of one activated sensor node.

Step 2: Upon receiving the node departure report, the sensor control server also adopts the local optimization Local_OPT on the subgraph to reschedule the sensing tasks for quality-of-sensing guaranteeing. Note that node x is excluded from the set S' in the subgraph. For example, as shown in Fig. 2.3 where sensor node 10 is departure, four targets of *Task* 2 and Task 3 become uncovered after the departure of 10. A subgraph consisting of nodes 6, 8 and 9 circled by the dashed quadrangle

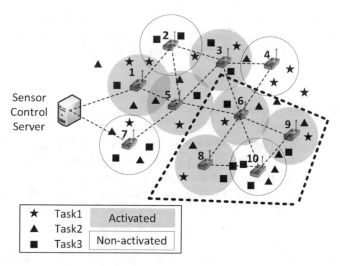

Fig. 2.3 New activation after the departure of node 10

will be established. Via applying Local_OPT to the subgraph, node 8 and 9 will be activated to ensure the coverage requirement.

Step 3: If the Local_OPT returns a feasible solution, the sensor control server will activate the sensors and assign sensing tasks to the activated sensors accordingly. To activate the sensor nodes in sleeping node, we can adopt the radio-triggered power management technique proposed in [34], where a sensor node can be activated via the power contained in activation message. Thus, negligible power is required. Otherwise, it uses a global optimization to find a feasible solution so as to guarantee the quality-of-sensing.

2.7 Performance Evaluation

In this section, we present our simulation-based performance evaluation results on the efficiency of our proposed online algorithm. We consider a 100×100 sensor network area randomly deployed with a number of software defined sensor nodes. The default number of sensor nodes is set as 400. Unless otherwise stated, these tasks are developed with 100 targets for each randomly distributed within the network. The corresponding program sizes for the three tasks are normalized as 0.3, 0.4, and 0.4. Both sensing rate requirement and duration are uniformly distributed for all tasks, with default expectation 100 Hz and 1 ms, respectively. The power consumption of a sensor in activated mode and sleeping mode is set as $P_a = 12.0$ mW and $P_s = 270\,\mu$W, respectively, which have been verified and adopted in [11]. The default sensing range is set as 6 m. The online algorithm is implemented in our simulator, where node participation and departure events are randomly generated at runtime, and each triggers both "local" and "global" results obtained by our online algorithm global optimization, respectively. All the MILPs are solved using commercial solver Gurobi optimizer [36].

2.7.1 Effective Sensing Rate

Before reporting our performance evaluation results, we first present the experimental results on the effective sensing rate that we can derive in Theorem 1. In the simulation, for a specific target of task t, the number of effective sensing events is counted under a certain overall sensing rate f. We first notice that the same result is achieved for the same overall sensing rate, no matter how it is distributed among the collaborative sensors. Such observation is consistent with the conclusion of compound Poisson process. Furthermore, Fig. 2.4a shows that the effective sensing rate increasing sublinearly with the overall sensing rate. This is because higher overall sensing rate indicates higher probability of duplicated sensing, under the same sensing duration. Figure 2.4b shows the effective sensing rate as a non-linear decreasing function of the sensing duration. When $c_t = 0$, the effective sensing rate equals to the overall sensing rate as all the sensing operations are effective, while when $c_t f = 1$, the effective sensing ratio drops as low as 50%, e.g., around 5.0 when $f = 10$ Hz. For the similar reason, longer sensing duration implies more duplicated sensing under the same sensing rate and hence lower effective sensing rate.

2.7.2 Rescheduling Time

To study the efficiency of our proposed online algorithm, we consider a SDSN randomly deployed with 400 sensor nodes. After the deployment, we simulate

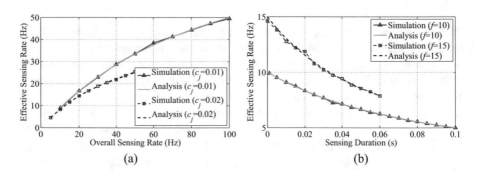

Fig. 2.4 Validation on the correctness and accuracy of Theorem 1. (**a**) On the overall sensing rate. (**b**) On the sensing duration

the sensor node dynamics as a Poisson process with average rate 0.1 events per time unit. The rescheduling time (i.e., the calculation time for the new solution) is investigated on a server configured with i7 quad-core 3.4 GHz CPU, 8 GB memory, and Python 2.7.5. As a case study, Fig. 2.5a shows the instant rescheduling time for each dynamic event along 1000 time units. Obviously, in most cases, "local" requires much lower rescheduling time, compared to "global" one. Recall that our proposed online algorithm requires only local network information while the global one always asks for the information of the entire network. As a result, although both global algorithm and our local algorithm are with similar formulation, the smaller input (i.e., subgraph) makes our proposed online algorithm have much lower computation complexity than the global one requiring full graph information. However, in few cases, we also notice that "local" has longer rescheduling time than "global" one. For example, at time 152.82, the rescheduling time of "global" and "local" is 21.1 s and 21.6 s, respectively. This is because we cannot find a feasible solution via "local" optimization and have to resort to "global" one. Longer rescheduling time is thus incurred. Furthermore, one may also notice that the rescheduling time is not stable along the time line. The main reason is that the number of involving sensors and targets varies quite much from case to case in all randomly generated node participation/departure events. To clearly compare the rescheduling time of our local optimization and the global one, we further plot the cumulative distribution function (CDF) of the rescheduling time in Fig. 2.5b. Figure 2.5b gives the CDF of the instant rescheduling time shown in Fig. 2.5a. We can see that "local" exhibits outstanding rescheduling time advantage over the "global." For example, with probability around 60%, "local" can achieve a rescheduling time less than 0.323 s while the one for "global" is 18.78 s. According to the above performance evaluation results, we can conclude that our online algorithm indeed outperforms global optimization in rescheduling computation complexity.

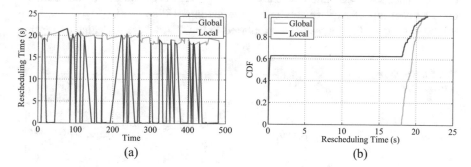

Fig. 2.5 On the rescheduling time. (**a**) Instant rescheduling time for 1000 time units. (**b**) CDF of the rescheduling time

2.7.3 Power Efficiency

Although our online algorithm has advantages on the control efficiency, one may naturally wander how the "Local" algorithm performs on the network power efficiency. In this section, we vary different parameters such as the number of sensor nodes, the coverage ratio requirement, the transmission range to extensively check the power efficiency of our proposed algorithm. Before any dynamic event, "Original" results are obtained as references.

We first investigate the effect of network size $|S|$ under the values of $c_t = 1\,\text{ms}, f_t = 100\,\text{Hz}, \delta_t = 0.3, r_t = 6, \forall t \in T$. The number of sensors $|S|$ is varied from 150 to 500. Figure 2.6 shows the results after sensor node departure event. For all algorithms, it can be first noticed that the power consumption first decreases and then slightly increases network size $|S|$. This is because when the network size is small, more sensor nodes imply more candidates to satisfy the quality-of-sensing requirement and thus less power will be consumed. However, when the network size is large enough to easily find the sensors that shall be activated for quality-of-sensing guaranteeing, further increasing the network size does not benefit the power efficiency but introduces more non-negligible power consumption from sleeping nodes. Thus, the power consumption slightly increases. One may also observe that the gap between "local" and "global" shrinks with the increasing of network size. The reason is that there are more candidate sensors near the departure sensor node when the sensor node density is high and it is easy to find a substitution to the departure one. Actually, we find out that the "local" result sometimes is even the same as "global" one in some cases. In few cases, two or more sensor nodes are needed to compensate the departure of one sensor node with many sensing tasks. Nevertheless, under any network size, we can always see the closeness of our "local" algorithm to the optimal "global" one.

According to Theorem 1, the sensing rate requirement also has deep influence on the sensor activation and hence the overall power consumption. To this end, we conduct a group of experiments under the settings $|S| = 400, c_t = 1\,\text{ms}, \delta_t = 0.3, r_t = 6, \forall t \in T$. The results are reported in Fig. 2.7, where the value of sensing

Fig. 2.6 On the effect of network size

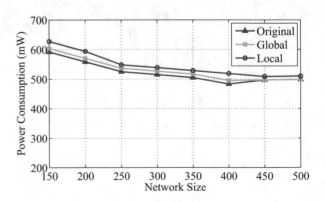

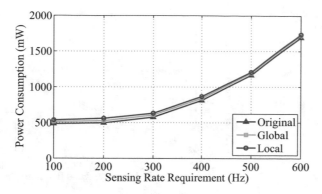

Fig. 2.7 On the effect of sensing rate requirement

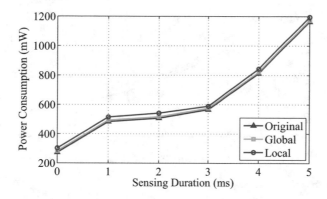

Fig. 2.8 On the effect of sensing rate requirement

rate requirement varies from 100 to 600 Hz. We observe that the overall power consumption shows as a superlinear increasing function of the required sensing rate. When the rate is low, to activate a single sensor is sufficient to conduct the sensing tasks for many targets. When the rate is high, the number of activated sensors sharply increases in order to satisfy the tough sensing requirement.

Besides the sensing rate requirement, the sensing duration also affects the power consumption. To this end, we also investigate the effect of sensing duration. The experimental results, where the duration varying from 1 to 5 ms, are shown in Fig. 2.8 under the setting $|S| = 400$, $f_t = 100$ Hz, $\delta_t = 0.3$, $r_t = 6$, $\forall t \in T$. We notice that the power consumption is a superlinear function of the sensing duration for the similar reason.

Next, we check how the quality-of-sensing requirement affects the network power efficiency. Figure 2.9 shows the results under the values of $|S| = 500$, $c_t = 1$ ms, $f_t = 100$ Hz, $r_t = 6$, $\forall t \in T$. The coverage ratio requirement for all tasks is increased from 0.1 to 0.6. We find out that the power consumption almost linearly increases with the values of δ for all algorithms. This is due to the fact that more

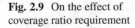

Fig. 2.9 On the effect of coverage ratio requirement

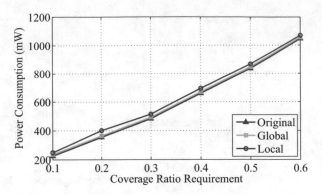

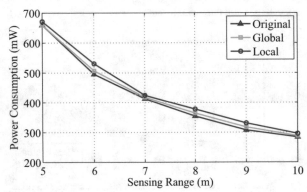

Fig. 2.10 On the effect of sensing range

sensors shall be activated to cover more targets to satisfy the required quantity-of-sensing. For example, according to the "original" results, around 32 nodes shall be activated when the coverage ratio is set as 0.3 while the value increases to 70 when the requirement becomes 0.6. After a dynamic event, "global" achieves almost the same performance as "original" while "local" requires little higher power as only local information is exploited. For example, as shown in Fig. 2.3, after the departure of node 10, node 7 can be activated to ensure the coverage if global optimization is applied, other than activating both nodes 8 and 9 by local optimization. However, remember that global optimization is at the expense of high scheduling time and control overhead.

We finally check the effect from the sensing range, which determines the reachability of sensor nodes to the targets. We investigate its effect via setting $|S| = 500, c_t = 1\,\text{ms}, f_t = 100\,\text{Hz}, \delta_t = 0.3, \forall t \in T$. and vary the sensing range from 5 to 10 m. The evaluation results are shown in Fig. 2.10, from which we can see that the power consumption shows as a decreasing function of the sensing range. This is consistent with our intuition that more targets can be reached by the sensor nodes with longer sensing duration and potentially smaller number of sensor nodes shall be activated to ensure the quality-of-sensing. However, we further notice that the decreasing rate becomes low with the increasing of sensing range. This is because reachability is not equivalent to coverage. Although more targets can be reached with long sensing range, the ability to cover these targets is still constrained by the schedulability and resource capacities. Thus, further increasing of sensing range does not take too much benefit to the power efficiency any more.

2.8   Conclusion

In this chapter, we consider a minimum-power activation and scheduling problem in multi-task SDSNs with quality-of-sensing guaranteed. We first formally derive the effective sensing rate that can be achieved by collaborative sensing from multiple sensors in closed-form. Based on our analysis, we build an MIQP formulation to describe the minimum-power activation problem and then linearize it into MILP to lower the computation complexity. We further notice that it is unnecessary to always apply global optimization on the whole network. To this end, we further propose an online algorithm using local information near the dynamic event point to deal with the dynamic events that may happen during the SDSN runtime. Through extensive simulation studies, we prove the high efficiency our online algorithm using local optimization by the fact that it much approaches the network power efficiency using global optimization but substantially outperforms it on rescheduling time.

Chapter 3
Software Defined Communication

Abstract The fast development of mobile computing has raised ever-increasing diverse communication needs in wireless networks. To catch up with such needs, cloud-radio access networks (CRAN) is proposed to enable efficient radio resource sharing and management. By CRAN, it is possible to realize software defined access networks. At the same time, the massive deployment of radio access networks has caused huge energy consumption. Incorporating renewable green energy to lower the brown energy consumption also has become a widely concerned topic. In this chapter, we are motivated to investigate a green energy aware remote radio head (RRH) activation problem for coordinated multi-point (CoMP) communications in green energy powered CRAN, aiming at minimizing the network brown energy power consumption. The problem is first formulated into a non-convex optimization form. By analyzing the characteristics of the formulation, we further propose a heuristic algorithm based on an ordered selection method. Extensive simulation based experiment results show that the proposed green energy aware algorithm provides an effective way to reduce brown energy power consumption, well fitting the goal of developing green communications.

3.1 Background on Cloud-Radio Access Networks

To pursue the vision of smart cities, a massive number of wireless devices (e.g., smart meters, various sensors, intelligent transportation devices, etc.) have been penetrated into the cities during the past decades. This results in an exponential growth in both the number of connected user equipments (UE) and the volume of data. According to the report from Ericsson, the number of mobile subscriptions will reach 7.7 billion by 2021. Due to such fact, it is estimated that the total mobile data traffic will reach 30.6 Exabytes by 2020, according to the report from Cicso. To satisfy such huge mobile subscriptions and data traffic, a lot of base stations shall be deployed. The vast deployment of base stations results in huge energy consumption, raising wide concern among the mobile operators [72]. Such trend imposes challenges to both the performance efficiency and power efficiency of radio networks.

D. Zeng et al., *Software Defined Systems*, SpringerBriefs in Computer Science,
https://doi.org/10.1007/978-3-030-32942-6_3

To cope with these challenges, a newly emerging technology, named cloud-radio access networks (CRAN), has been proposed and shown great potential in promoting the performance efficiency, energy efficiency as well as the network flexibility of future wireless networks. CRAN decouples the baseband processing function from traditional base station and centralizes it into shared baseband unit (BBU) pool, by exploring technologies like network function virtualization (NFV) and software defined network (SDN) [48–50, 119]. Via embracing edge computing technology [35, 88, 89, 106, 107, 126], BBU pool can also reside in the edge cloud, referred as Fog Radio Access Networks (FRAN). By such means, the front-end becomes simple and lightweight remote radio heads (RRHs) that can be densely deployed to provide radio access services for different UEs, well fitting the fast growing communication demands of smart city applications.

Meanwhile, another developing trend on the wireless networks is on the revolution of power provision paradigm, i.e., from traditional non-renewable brown energy to renewable green energy. By powering wireless networks via distributed renewable resources (DERs) that harvest green energy from the environment (e.g., solar, geothermal, wind, tide, and hydro), it has shown great potential in pursuing high network energy efficiency [60, 80]. By decoupling the baseband processing, the front-end RRHs become simple and lightweight. Therefore, it is natural to power the RRHs with green energy from DERs. This motivates us to consider a green energy powered CRAN architecture, as shown in Fig. 3.1, where all the RRHs are powered by both green energy from DERs and brown energy from the power grid [127]. In such architecture, how to manage the network resources with the consideration of green energy generation condition therefore becomes a critical issue to be tackled.

3.2   RRH Activation Problem

CRAN has been widely regarded as an ideal platform to realize coordinated multi-point (CoMP) communications. By centralizing the baseband processing, a BBU can compute the beamforming weight coefficients for different RRHs that serve one UE. Upon the reception of the precoded data, the RRHs then can cooperatively transmit the data to the served UE, which shall then observe the superposition of multiple signals from the serving RRHs. By coherently combining the signals, a high signal-to-interference-plus-noise ratio (SINR) shall be achieved, potentially improving the network's spectral efficiency and performance efficiency. In order to save energy consumption, it has been shown that not all, but a subset of, RRHs need to be activated, provided that the predefined Quality of Service (QoS) is guaranteed. Although energy efficient RRH activation has been widely discussed in the literature, e.g., [100, 101], we notice that none of the existing studies takes the green energy generation characteristics into consideration. With the consideration of green energy, aggressively shutting down the RRHs to minimize the number of RRHs activated does not always mean high energy efficiency. Instead, we shall carefully choose the RRHs that shall be activated to form a CoMP cluster for

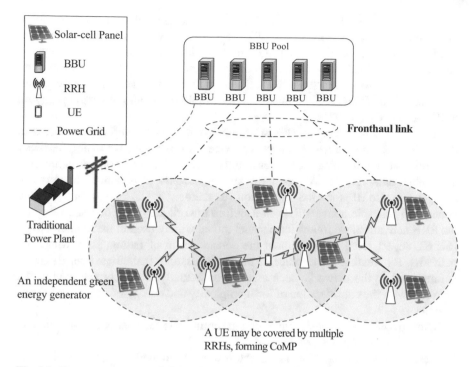

Fig. 3.1 Green energy powered C-RAN architecture. All the RRHs are powered by both legacy power grid and an independent green energy generator. A UE may be covered by multiple RRHs, forming CoMP

each UE. This motivates us to investigate the green energy aware RRH activation for CoMP communications in green energy powered CRAN in this chapter. We first formally describe the problem into a non-convex optimization programming problem. Then, by reformulating and relaxing the original problem into a convex one, we further propose a heuristic green energy aware RRHs activation algorithm. Through extensive simulations, the high energy efficiency of our green energy aware RRHs selection algorithm is verified by the fact that it indeed outperforms the competitor without the consideration of green energy generation characteristics.

3.2.1   System Model

We consider a CRAN, as shown in Fig. 3.1, with a set $\mathcal{L} = \{1, 2, \cdots, L\}$ of RRHs randomly scattered in the network, where RRH $l \in \mathcal{L}$ is equipped with N_l antennas. All the RRHs are simultaneously powered by legacy power grid with brown energy and DER with renewable green energy. We assume that the average green energy generation rate at DER on RRH l is $P_{g,l}$. Due to the separation of RRH and BBU, the fronthaul network connecting RRHs and BBUs requires an extremely high bandwidth to transport the digital I-Q samples [2]. Usually, high-speed medium is adopted by the fronthaul network. It could be in either wired or wireless way, e.g., fiber or microwave. We assume that a fixed amount of fronthaul network capacity is shared by the RRHs and the maximum allowable fronthaul network capacity on RRH $l \in \mathcal{L}$ is $C_{l,\max}$ [37].

A set $\mathcal{K}$ of single-antenna UEs are randomly distributed and are connected to these RRHs to acquire network access service. In this paper, we mainly consider downlink data communications. To ensure the QoS, UE $k \in \mathcal{K}$ requires a minimum achievable data rate r_k. As we have known, CRAN is a natural platform to easily adopt CoMP for QoS promotion. Therefore, a UE may simultaneously connect to multiple active RRHs to acquire network service. For traceability on the RRH activation, we follow the channel model widely used in the literature (e.g., [61, 62, 86, 91, 105, 114]), without the consideration of fading, shadowing, and path loss. By treating interference as noise, we assume that single user detection is employed in the network. Such assumption actually is widely accepted in the literature with the consideration of low-complexity and energy-efficient structure of antenna.

In this paper, we are mainly interested in minimizing the energy consumption on the RRHs, which consists of three parts.

The first part is for the wireless signal transmission to the UEs and is proportional to the beamforming vector, i.e., $P_l^{tr} = \sum_{k \in \mathcal{K}} \| \mathbf{w}_{lk} \|_{\ell_2}^2$. It is widely noted that a RRH $l \in \mathcal{L}$ is with a predefined transmission power limitation P_l that cannot be surpassed, regardless of the usage of brown energy or green energy.

A high-capacity fronthaul connection is established between a RRH and its associated BBU. It is noticed that the power consumption on the high-speed fronthaul links is comparable to the one for wireless transmissions [108]. Therefore, the fronthaul network energy consumption is un-ignorable. As each activated RRH must associate with its BBU via one fronthaul link, we calculate the fronthaul energy consumption due to RRH l as $P_l^{fr} = \sum_{k \in \mathcal{K}} \frac{1}{\eta_l} \| \mathbf{w}_{lk} \|_{\ell_2}^2$, where η_l is the drain efficiency of the radio frequency power amplifier on RRH l [3].

Besides the above two kinds of energy consumption, another un-ignorable part is known as the static energy consumption, independent of the signal received and sent. Whenever a RRH $l \in \mathcal{R}$ is activated, a fixed amount P_l^c of energy, mainly depending on the number of antennas, is consumed.

As a result, we can express the energy consumption on a RRH $l \in \mathcal{L}$ as:

$$P_l = \sum_{k \in \mathcal{K}} \| \mathbf{w}_{lk} \|_{\ell_2}^2 + \sum_{k \in \mathcal{K}} \frac{1}{\eta_l} \| \mathbf{w}_{lk} \|_{\ell_2}^2 + P_l^c, \forall l \in \mathcal{L}. \tag{3.1}$$

3.2.2 Problem Statement and Formulation

With respect to energy efficiency, there is no need to activate all RRHs provided that the predefined QoS of all users get satisfied. Intuitively, the less RRHs activated, the higher energy efficiency can be achieved. However, by taking the green energy into consideration, we are mainly interested in maximizing the usage of green energy so as to minimize the brown energy consumption. Appropriate selection of the active RRH set with the consideration of diverse green energy generation rates is critical to the energy efficiency of the network. Besides, as the RRHs cooperatively send the signal with different weights to the users. It is also of great importance to appropriately set the weight on each RRHs for a UE. As a result, our problem can be stated as: how to appropriately choose a subset $\mathcal{A} \subseteq \mathcal{L}$ and set the weight on the active RRHs for different users, so as to minimize the brown energy consumption, while still guaranteeing the predefined QoS of all users.

QoS Constraints By enabling CoMP, a UE may associate with multiple active RRHs at the same time. With respect to energy efficiency, not all RRHs need to be active but a subset $\mathcal{A} \subseteq \mathcal{L}$ shall be activated provided the predefined QoS requirements of all users get satisfied. Hence, the baseband received signal at UE $k \in \mathcal{K}$ is given by

$$y_k = \sum_{l \in \mathcal{A}} \mathbf{h}_{kl}^{H} \mathbf{w}_{lk} s_k + \sum_{i \neq k} \sum_{l \in \mathcal{A}} \mathbf{h}_{kl}^{H} \mathbf{w}_{li} s_i + z_k, \forall k \in \mathcal{K}, \tag{3.2}$$

where s_k is a complex scalar denoting the data symbol for user k, $\mathbf{w}_{lk} \in \mathbb{C}^{N_l}$ is the beamforming vector at RRH l for user k, $\mathbf{h}_{kl} \in \mathbb{C}^{N_l}$ is the channel state information (CSI) vector from RRH l to user k, and $z_k \in \mathcal{CN}(0, \sigma_k^2)$ is the additive Gaussian noise.

The corresponding signal-to-interference-plus-noise ratio (SINR) for UE k is hence given by

$$\text{SINR}_k = \frac{\left| \sum_{l \in \mathcal{A}} \mathbf{h}_{kl}^{H} \mathbf{w}_{lk} \right|^2}{\sum_{i \neq k} \left| \sum_{l \in \mathcal{A}} \mathbf{h}_{kl}^{H} \mathbf{w}_{li} \right|^2 + \sigma_k^2}, \forall k \in \mathcal{K}. \tag{3.3}$$

Note that we theoretically consider that a UE can be served by any RRH in (3.3).

In order to ensure the minimum achievable data rate of UE $k \in \mathcal{K}$, according to the Shannon's theorem, the SINR value of UE k shall be larger than a predefined value γ_k [101]. That is,

$$\text{SINR}_k \geq \gamma_k, \forall k \in \mathcal{K}. \tag{3.4}$$

RRH Transmission Power Constraints As we have known, the maximum transmission power consumption each RRH that can tolerate is limited. Therefore, the total transmission power consumption for all users on a RRH is constrained by the maximum allowable power P_l, i.e.,

$$\sum_{k \in \mathcal{K}} \| \mathbf{w}_{lk} \|_{\ell_2}^2 \leq P_l, \forall l \in \mathcal{A}. \tag{3.5}$$

Fronthaul Network Capacity Constraints The fronthaul link is used to carry the signal from the BBUs to RRHs. For each UE $k \in \mathcal{K}$, the fronthaul network capacity consumption on a RRH is proportional to its data rate, i.e., the number of data symbols to be carried by the fronthaul link from BBU to RRH l for UE k. By adopting CoMP, a number of RRHs may cooperate with each other to serve one user. If a UE k is served by RRH $l \in \mathcal{L}$, it indicates that its corresponding beamforming vector $\mathbf{w}_k$ is nonzero. To this end, we define an indicator function as:

$$f \left(\| \mathbf{w}_{lk} \|_{\ell_2}^2 \right) = \begin{cases} 0, & \text{if } \| \mathbf{w}_{lk} \|_{\ell_2}^2 = 0, \\ 1, & \text{if otherwise.} \end{cases} \tag{3.6}$$

Now, following [19], we can describe the fronthaul network bandwidth requirement of an active RRH $l \in \mathcal{A}$ as the accumulated data rates to be transmitted between the BBU pool and RRH l for all UEs, i.e.,

$$C_l = \sum_{k \in \mathcal{K}} f \left(\| \mathbf{w}_{lk} \|_{\ell_2}^2 \right) \cdot r_k, \forall l \in \mathcal{A}. \tag{3.7}$$

As we have known, the fronthaul network capacity for each RRH is limited. As a result, the total bandwidth requirement on a RRH shall not exceed its capacity, i.e.,

$$\sum_{k \in \mathcal{K}} f \left(\| \mathbf{w}_{lk} \|_{\ell_2}^2 \right) \cdot r_k \leq C_{l,\max}, \forall l \in \mathcal{A}. \tag{3.8}$$

Problem Formulation The RRHs are simultaneously powered by brown energy from legacy power grid and green energy from DERs. Our objective to maximize the energy efficiency is equivalent to minimizing the brown energy usage. For each RRH $l \in \mathcal{A}$, the brown energy consumption $P_{b,l}$ can be calculated as:

$$P_{b,l} = \max\{0, P_l - P_{g,l}\}$$

$$= \max \left\{ 0, \sum_{k \in \mathcal{K}} \left(1 + \frac{1}{\eta_l} \right) \| \mathbf{w}_{lk} \|_{\ell_2}^2 + P_l^c - P_{g,l} \right\}, \forall l \in \mathcal{A}, \tag{3.9}$$

by which we can express the total brown energy consumption on all RRHs as:

$$p_b(\mathcal{A}, \mathbf{w}) = \sum_{l \in \mathcal{A}} \max \left\{ 0, \sum_{k \in \mathcal{K}} \left(1 + \frac{1}{\eta_l} \right) \| \mathbf{w}_{lk} \|_{\ell_2}^2 + P_l^c - P_{g,l} \right\},$$

where $\mathbf{w} = \left[\mathbf{w}_{11}^T, \dots, \mathbf{w}_{1k}^T, \dots, \mathbf{w}_{L1}^T, \dots, \mathbf{w}_{LK}^T \right]^T$.

Now, we can formulate the brown energy consumption problem as:

$$\min : p_b(\mathcal{A}, \mathbf{w})$$

$$\text{s.t.} \quad (3.4), (3.5), (3.8).$$

It can be seen that problem $\mathscr{P}$ is with variables $\mathbf{w}$ and $\mathcal{A}$, referring to the active RRH set selection and the beamforming weight settings, respectively. It is an NP-hard non-convex problem and is difficult to solve in general. We will further analyze it and try to reformulate it in the next section.

3.2.3 Problem Reformulation and Analysis

We notice that the objective function, transmission power, and fronthaul network capacity constraints are all not affected by any phase rotation of the beamforming vector, i.e., $\mathbf{w}_k = [\mathbf{w}_{1k}^T, \ldots, \mathbf{w}_{lk}^T, \ldots, \mathbf{w}_{Lk}^T]^T \in \mathbb{C}^{\sum_{l\in\mathcal{A}} N_l}$. As a result, according to [116], we can rewrite the QoS constraint for user $k \in \mathcal{K}$ in a second-order cone (SOC) form as:

$$C_1(\mathcal{A}, \mathbf{w}) : \sqrt{\sum_{i\neq k} |\mathbf{h}_k^H \mathbf{w}_i|^2 + \sigma_k^2} \leq \frac{1}{\sqrt{\gamma_k}} |\mathbf{h}_k^H \mathbf{w}_k|^2, \forall k \in \mathcal{K}, \tag{3.10}$$

where $\mathbf{h}_k^H = [\mathbf{h}_{k1}^H, \ldots, \mathbf{h}_{kl}^H, \ldots, \mathbf{h}_{kL}^H] \in \mathbb{C}^{\sum_{l\in\mathcal{A}} N_l}$.

The RRH transmission power constraints can be also rewritten into SOC form as:

$$C_2(\mathcal{A}, \mathbf{w}) : \sqrt{\sum_{k\in\mathcal{K}} \|\mathbf{w}_{lk}\|_{\ell_2}^2} \leq \sqrt{P_c^l}, \forall l \in \mathcal{A}. \tag{3.11}$$

For the fronthaul network capacity constraints, we first equivalently rewrite the indicator function $f(\|\mathbf{w}_{lk}\|_{\ell_2}^2)$ as:

$$f\left(\|\mathbf{w}_{lk}\|_{\ell_2}^2\right) = \|\|\mathbf{w}_{lk}\|_{\ell_2}^2\|_{\ell_0}. \tag{3.12}$$

According to [10], we can approximate the non-convex ℓ_0-norm expressions by a convex ℓ_1-norm to deal with the discrete indicator function in constraints (3.8) as follows:

$$\|\mathbf{w}\|_{\ell_0} \approx \sum_{i\in\mathcal{K}} \beta_i |w_i|, \tag{3.13}$$

where w_i is the i-th element in vector $\mathbf{w}$ with weight β_i. Then, following [19], the constraints (3.8) in $\mathscr{P}$ can be rewritten as:

$$C_3(\mathcal{A}, \mathbf{w}) : \sum_{k \in \mathcal{K}} \beta_{lk} \|\mathbf{w}_{lk}\|_{\ell_2}^2 \cdot r_k \leq C_{l,\max}, \forall l \in \mathcal{A}, \tag{3.14}$$

where

$$\beta_{lk} = \frac{1}{\|\mathbf{w}_{lk}\|_{\ell_2}^2 + \tau}, \forall k \in \mathcal{K}, l \in \mathcal{A} \tag{3.15}$$

and τ is a small positive factor to ensure stability and can be set as $\tau = 10^{-10}$ [19].

Now, problem $\mathcal{P}$ can be transformed into

$$\mathcal{P}_1 : \quad \min_{\mathbf{w}, \mathcal{A}} \quad p_b(\mathcal{A}, \mathbf{w})$$

$$\text{s.t.} \quad C_1(\mathcal{A}, \mathbf{w}), C_2(\mathcal{A}, \mathbf{w}), C_3(\mathcal{A}, \mathbf{w}).$$

Similarly, $\mathcal{A}$ and $\mathbf{w}$ are the variables to be derived. Let us first consider the case where the active RRH set $\mathcal{A}$ is given and fixed. This shall result in a brown energy consumption minimization problem $\mathcal{P}_1(\mathcal{A})$. Thereafter, we can search over all the possible active RRH sets based on the solution of $\mathcal{P}_1(\mathcal{A})$ in a brute-force manner. Therefore, we have

$$p_{opt} = \min_{\mathcal{A}^* \in \{\mathcal{A}_1, \dots, \mathcal{A}_n\}} p_{opt}(\mathcal{A}^*), \tag{3.16}$$

where $p_{opt}(\mathcal{A}^*)$ is the optimal value of the problem $\mathcal{P}_1(\mathcal{A}^*)$. $\{\mathcal{A}_1, \dots, \mathcal{A}_n\}$ is the set of fixed RRH sets.

The number of set $\mathcal{A}^*$ for a given RRH set with cardinality m is $\binom{L}{m}$, which can be very large. Besides, note that the value of m could range between 1 and L in theory. As a result, the searching space discussed above shall exponentially increase with the number of RRHs L. Therefore, brute-force searching method is not feasible in practice. Therefore, we propose a heuristic algorithm in the next Section.

3.2.4 Green Energy Aware RRH Activation Algorithm Design

In order to avoid treating $\mathcal{A}$ as variables or brute-force searching of all the feasible active RRH set $\mathcal{A}$, we propose a heuristic green energy aware RRH activation algorithm in this Section.

By analyzing $\mathcal{P}_1$, we notice that $p_b(\mathcal{A}, \mathbf{w})$ is a convex function if the active RRH set $\mathcal{A}$ is given. However, even with given $\mathcal{A}$, it is still difficult to solve $\mathcal{P}_1$ due to the involvement of the fronthaul network capacity constraints C_3, where the value of r_k is also related with variables $\mathbf{w}$. To address this problem, we plan to further temporarily exclude the fronthaul network capacity constraints C_3. By such means, we can obtain a solvable SOC problem as:

$$\mathscr{P}_{soc}(\mathcal{A}): \quad \min_{\mathbf{w}} \quad p_b(\mathbf{w})$$

$$\text{s.t.} \quad C_1(\mathbf{w}), C_2(\mathbf{w}),$$

where $\mathbf{w}$ is the optimization variable to be derived for a given $\mathcal{A}$. The above observations motivate the main concept of our algorithm as: we first treat all RRHs as active and then try to iteratively switch off the RRHs with certain rule and solve $\mathscr{P}_{soc}(\mathcal{A})$ with the updated $\mathcal{A}$ until we find out a feasible solution that can minimize the brown energy consumption, without violating any constraints discussed in Sect. 3.3.2. Our proposed heuristic algorithm is summarized in Algorithm 1 and will be detailed as follows.

We first initialize $\mathcal{A}$ as the whole RRH set, i.e., $\mathcal{A} = \mathcal{L}$, and the inactive RRH set is accordingly initialized as empty, i.e., $\mathcal{D} = \emptyset$, as shown in line 1. We first solve $\mathscr{P}_{soc}(\mathcal{A})$ to obtain an initial solution on the beamforming vector $\mathbf{w}$ in line 2. If the problem is feasible, we then try to iteratively switch off RRHs in $\mathcal{A}$ until finding out a feasible solution that can minimize the brown energy consumption.

To this end, we propose a RRH ordering rule based on several factors (e.g., the beamforming vector already obtained, the fronthaul network capacity, RRH power consumption and the green energy generation rate, etc.) as:

$$\theta_l = \sqrt{\frac{\eta_l \sum_{k \in \mathcal{K}} \|\mathbf{h}_{kl}\|_{\ell_2}^2}{P_l^c + \sum_{k \in \mathcal{K}} \beta_{lk}} \left(\sum_{k \in \mathcal{K}} \|\mathbf{w}_{lk}\|_{\ell_2}^2 \right) + P_{g,l}}, \forall l \in \mathcal{L}, \qquad (3.17)$$

Algorithm 1 Green energy aware RRH activation algorithm towards energy efficient CoMP in CRAN

1: Initialization: the active RRH set $\mathcal{A} \leftarrow \mathcal{L}$, inactive RRH set $\mathcal{D} = \emptyset$
2: Take $\mathcal{A}$ into $\mathscr{P}_{soc}(\mathcal{A})$ and solve the convex optimization problem
3: **if** $\mathscr{P}_{soc}(\mathcal{A})$ is feasible **then**
4:      Obtain the corresponding beamforming vector $\mathbf{w}$
5:      Sort the RRHs in $\mathcal{A}$ in an ascending order $\theta_{\pi_1} \leq \ldots \leq \theta_{\pi_{|\mathcal{A}|}}$ according to the ordering criteria defined in (3.17)
6:      **repeat**
7:          Remove the first element, i.e., π_1, from active RRH set $\mathcal{A}$
8:          Solve $\mathscr{P}_{soc}(\mathcal{A})$ with updated $\mathcal{A}$
9:          **if** it is feasible **then**
10:              Insert π_1 into $\mathcal{D}$
11:         **end if**
12:     **until** $\mathscr{P}_{soc}(\mathcal{A})$ is infeasible
13:     Obtain the RRHs that shall be activated and the corresponding transmission beamformers for all users
14: **else if** $\mathscr{P}_{soc}$ is infeasible **then**
15:     go to **end**
16: **end if**
17: **end**

---

where the RRH with lower value of θ_l shall have higher priority to be switched off. It is obvious that θ_l is proportional to the channel power gain $\sum_{k\in\mathcal{K}} \|\mathbf{h}_{kl}\|_{\ell_2}^2$, which is related to the sum capacity of the whole network. The RRHs with higher channel power gain contribute more to the total capacity and therefore shall have lower priority to be switched off. Note that we should reconsider the fronthaul constraint $C_3(\mathbf{w})$, so we regard β_{lk} as one of the key system parameters to choose RRHs to be switched off. As for the heuristic weight updating rule (3.15), β_{lk} is inversely proportional to the transmit power level $\|\mathbf{w}_{lk}\|_{\ell_2}^2$. Switching off the RRH with lower transmit power, i.e., higher value of β_{lk}, shall have less impact on the QoS of the UEs, but more significant to the energy efficiency. Meanwhile, the fixed energy consumption P_c^l of an active RRH has similar effect as β_{lk}, i.e., a RRH with higher P_l^c should be encouraged to be switched off. Many existing studies, e.g., [43, 67, 129], show that a RRH l with small coefficient $r_l = (\sum_{k\in\mathcal{K}} \|\mathbf{w}_{kl}\|_{\ell_2}^2)^{1/2}$ contributes little beamforming gain and therefore shall be given high priority to be switched off. When green energy is take into consideration, one more important factor to the switch-off decision is the green energy generation rate $P_{g,l}$. There is no doubt that a RRH with higher green energy rate shall be encouraged to be active. Based on the above analysis, we design the ordering criteria as shown in (3.17).

After obtaining an initial active RRH set $\mathcal{A}$, we first sort the RRHs in an ascending order where the RRHs with lower θ_l shall have a higher priority to be switched off (line 5). According to (3.17), the RRHs in lower order shall have higher priority to be switched off. As a result, we then try to iteratively and greedily switch off RRHs until the problem $\mathscr{P}_{soc}(\mathcal{A})$ is infeasible any more, in order to minimize the brown energy as much as possible (lines 6–12). During each iteration, we first remove the RRH with the lowest value of θ_l from $\mathcal{A}$ and then try to solve $\mathscr{P}_{soc}(\mathcal{A})$ with the updated $\mathcal{A}$ to check its feasibility. When no RRH can be switched off any more, we terminate the iteration and finally obtain the RRHs that shall be activated, as well as their beamformers to all users (line 13).

Computational Complexity As shown in Algorithm 1, we shall iteratively solve the SOCP convex optimization problem $\mathscr{P}_{soc}(\mathcal{A})$ with computation complexity $\mathcal{O}((L-|\mathcal{D}|)^{3.5}N^{3.5}K^{3.5})$. Considering an extremely worst case, we need to go over all the RRHs, i.e., solving $\mathscr{P}_{soc}(\mathcal{A})$ with $L-1$ times. As a result, the proposed green energy aware RRH activation algorithm is with computation complexity $\mathcal{O}(L^{4.5}N^{3.5}K^{3.5})$.

3.2.5   Simulation Results

In this section, we present our simulation-based performance evaluation results to verify the correctness and efficiency of our proposed green energy aware algorithm (Aware). Specially, in order to show the advantage of our proposed algorithm, we compare the performance of our proposal against the state-of-the-art energy efficient RRH activation algorithm (Unaware) proposed in [100], which also intends to minimize the energy consumption but do not take the green energy generation rate into consideration. The unaware scheme also treats all RRHs as active to obtain the corresponding beamforming vector **w**. Next, it tries to deactivate the RRHs for energy saving by ordering the RRH according to the criteria related to the coefficient, channel power, and the transport link power consumption. For more details, please refer to reference [100].

Following the same environment settings as [100], we simulate a square region 1000×1000, where a number of RRHs and UEs are randomly distributed. Two different scales of networks are considered. In both network scales, the path and penetration loss is defined as $148 + 37.6 \log_2(d_{kl})$, where d_{kl} denotes the propagation distance between RRH l and UE k. The small scale fading is described as independent circularity symmetric complex Gaussian random variables with distribution $\mathcal{CN}(0, 1)$. The noise power spectral density is -102 dBm. The maximum allowable transmission power on any RRH is limited by $P_l = 1\,W, l \in \mathcal{L}$. All the transport link power consumption is $P_l^c = (5 + l)W, l \in \mathcal{L}$ and the power amplifier coefficient is 4. To verify the feasibility and efficiency of our work in large-scale network, we further consider large-scale networks where the number of users ranges from 20 to 50 and the number of RRHs also ranges from 20 to 50.

We investigate how our algorithm performs under different network settings, as well as how various parameters affect the brown energy power consumption, by varying the settings of SINR requirement SINR_k, the number of RRHs L and the number of UEs K. For each group of experiments, we compare the two algorithms under different values of average green energy generation rate $\bar{P}_g = \frac{\sum_{l \in \mathcal{L}} P_{g,l}}{|\mathcal{L}|}$, i.e., $\bar{P}_g = 0, 5$, and 10, and run the simulation for 50 times to obtain the average brown energy consumption.

We first vary the value of SINR requirements for all users from 0 to 6 dB and report the average brown energy consumption with different SINR requirements in Fig. 3.2. From both Fig. 3.2a on small-scale network and Fig. 3.2b on large-scale network, we can see that the brown energy consumption shows as an increasing function of the SINR requirement. This is due to the fact that higher transmit power is needed to guarantee the higher QoS requirement, definitely resulting in higher

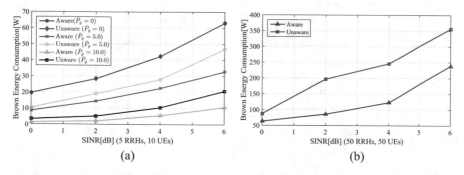

Fig. 3.2 Brown energy consumption on different SINR requirements. (**a**) Small-scale network. (**b**) Large-scale network

transmit power. And, if the green energy is not enough to satisfy such needs, we need to use the brown energy from the power grid, incurring the increasing of brown energy consumption. While, thanks to the efficient use of green energy, our proposed algorithm always requires less brown energy than the unaware algorithm, under any value of SINR requirement. Furthermore, by checking the brown energy consumption under different values of green energy generation rates $\bar{P}_g$, we can see that the brown energy consumption decreases with the increasing of green energy generation. Specially, we can see that the two curves for our "aware" algorithm and the "unaware" competitor overlap when $\bar{P}_g = 0$, this further verify that our algorithm can well adapt to the green energy generation rates and still performs as well as the state-of-the-art energy efficient RRH activation algorithm in [100].

Next, we investigate the energy efficiency of the two algorithms under different number of RRHs varying from 3 to 9 when the number of UEs is fixed as 10. All the UEs are with the same SINR requirement as 4 dB. Figure 3.3 presents the brown energy consumption on different number of RRHs. Similarly, the results for both small-scale and large-scale networks are shown. In either network scale, we notice that the brown energy consumption decreases with the increase of the number of RRHs. This is because we may have more feasible CoMP structure that can guarantee the QoS of all UEs when there are more RRHs, potentially enlarging the solution space leading to lower brown energy consumption. Once again, we notice that when $\bar{P}_g = 0$, the two algorithms show similar performance. While, in other cases, we can always observe the higher energy efficiency of our proposed green energy "aware" algorithm.

Finally, we vary the number of UEs in the network to investigate how the brown energy consumption is influenced by the number of UEs. We set the number of RRHs as 5 and 50 in small-scale network and large-scale network, respectively. The SINR requirements on all UEs are set as 4 dB. The performance evaluation results are reported in Fig. 3.4, from which we notice that the brown energy consumption shows as an increasing function of the number of UEs. This is because more users indicate more QoS requirements to be satisfied and henceforth higher energy

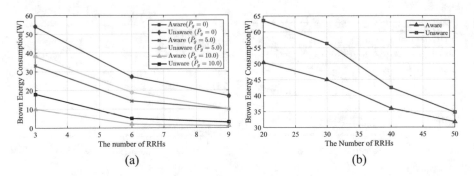

Fig. 3.3 Brown energy consumption on different number of RRHs. (**a**) Small-scale network. (**b**) Large-scale network

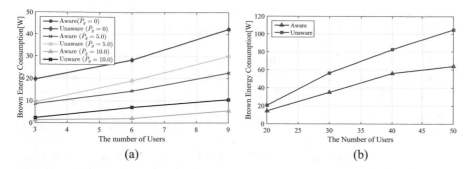

Fig. 3.4 Brown energy consumption on different number of users. (**a**) Small-scale network. (**b**) Large-scale network

consumption is needed. When the green energy fails to satisfy such needs, more brown energy consumption is needed. Nevertheless, the high energy efficiency of our proposal algorithm can be observed once again in Fig. 3.4. This further validates the high efficiency of our proposed algorithm.

3.3   Baseband Processing Unit Management

As we have seen, green energy aware network resource management is essential to promote the network energy efficiency. Besides RRHs, BBUs also contribute considerable amount of energy consumption. Therefore, we shall also discuss how to optimize the BBU energy efficiency. In this chapter, we assume that both RRHs and BBUs in C-RAN are powered simultaneously by both brown energy and green energy. The green energy can be transferred from one site to another, with certain energy attenuation. As a result, it is significant to study how to schedule the energy sharing together with the network resource allocation to minimize the brown energy usage. To this end, we study on the joint optimization of RRH-BBU association and green energy sharing. We formulate the joint optimization problem aiming at brown energy minimizations as an MILP problem and propose an efficient two-phase heuristic algorithm in this section.

3.3.1   System Model

We consider a general renewable-energy powered C-RAN network as shown in Fig. 3.1. The main network entities in a C-RAN network are the BBUs and RRHs, which are denoted as set B and R, respectively. The BBUs and RRHs are interconnected and scattered in the network area. The RRHs directly serve the users within its transmission area. Let λ_r denote communication request arrival rate on

RRH $r \in R$. Each BBU can process a limited amount of workload constrained by its processing capacity μ_{max}. We assume that all BBUs are homogeneous with the same processing capacity.

We adopt an energy consumption model given in [60] where each transmission consumes a unit of energy from the base station. Accordingly, we calculate the total energy consumed at any active BBU $b \in B$ as:

$$p_b = e_b + \alpha \mu_b, \tag{3.18}$$

where e_b is the static energy consumption of BBU b, regardless of its actual usage once activated and $\alpha \mu_b$ is the dynamic part proportional to the workload μ_b with ratio α. Similarly, the power consumed by RRH $r \in R$ can be calculated as:

$$p_r = e_r + \beta \lambda_r, \tag{3.19}$$

where e_r is the static energy consumption of RRH r, and $\beta \gamma_r$ is the dynamic part proportional to the workload λ_r with ratio β.

We assume that all the network entities are attached with a DER such that they are simultaneously powered by both brown energy from traditional power grid and the green energy from DER. The power generated at a DER is typically random with certain rate. Without loss of generality, we assume that the energy generated from DER at network entity $b \in B$ and $r \in R$ is with rate g_b and g_r, respectively. The green energy can be shared among the network entities. As indicated in [131], it is impossible to transfer power between DERs without any loss. The loss may account for 7% of the transferred energy or may even reach 55% in extreme cases. We use η_{xy}, $x, y \in R \cup B$ to denote the efficiently energy transferring ratio between the DERs on different network entities. Specially, we define $\eta_{xx} \equiv 1, \forall x \in R \cup B$ to indicate that no energy loss will be experienced at local DER.

3.3.2 Problem Formulation

RRH-BBU Association Constraints Thanks to the pooling of BBU resources, the association between BBU and RRH becomes flexible and can be adjusted at runtime. We denote the association relationship between a RRH $r \in R$ and a BBU $b \in B$ as a binary variable

$$x_{rb} = \begin{cases} 1, & \text{if RRH } r \text{ is associated with BBU } b, \\ 0, & \text{otherwise.} \end{cases} \tag{3.20}$$

Each RRH must be attached to one BBU, i.e.,

$$\sum_{b \in B} x_{rb} = 1, \forall r \in R. \tag{3.21}$$

Note that, from the consideration of energy efficiency, not all BBUs must be activated. To denote such status, we define a binary variable

$$y_b = \begin{cases} 1, & \text{if BBU } b \text{ is active,} \\ 0, & \text{otherwise.} \end{cases} \tag{3.22}$$

Consequently, the value of x_{rb} shall be constrained by y_b as:

$$x_{rb} \leq y_b, \forall r \in R, b \in B, \tag{3.23}$$

indicating that a RRH can only be associated with an active BBU, i.e., $y_b = 1$; otherwise, x_{rb} is forced to be 0 if $y_b = 0$.

Process Capacity Constraints According to the C-RAN communication process, the workload on a BBU is related to its associated RRHs. The total workload $\mu_b = \sum_{r \in R} x_{rb} \lambda_r$ allocated to BBU $b \in B$ shall not exceed its processing capacity, i.e.,

$$\sum_{r \in R} x_{rb} \lambda_r \leq y_b \mu_{\max}, \forall b \in B. \tag{3.24}$$

Power Supply Constraints To describe the energy transferring relationship, we define variables $p_{xy}, x, y \in R \cup B$ to represent the energy transferring amount from DER at network entity x to y. Specially, note that we allow $x = y$, e.g., $p_{xx}, x \in R \cup B$, indicating the amount of energy from the local DER. When the green energy cannot catch up with the energy demand of a network entity, certain brown energy must be retrieved from the power grid. We use p_{ey} to denote the amount of brown energy retrieved from the power grid to the network entity y, which could be either a RRH or a BBU. To ensure the functionality of a network entity, its total energy supply must satisfy its energy requirement. That is,

$$\sum_{r \in R} p_{rb} \eta_{rb} + \sum_{x \in B} p_{xb} \eta_{xb} + p_{eb} \geq y_b e_b + \alpha \sum_{r \in R} x_{rb} \lambda_r, \forall b \in B, \tag{3.25}$$

and

$$\sum_{x \in R} p_{xr} \eta_{xr} + \sum_{b \in B} p_{br} \eta_{br} + p_{er} \geq e_r + \beta \lambda_r, \forall r \in R, \tag{3.26}$$

where p_{eb} and p_{er} denote the brown energy retrieved from the power grid to BBU b and RRH r, respectively.

The green energy transferring amount is limited by the green energy generation rate. Therefore, we have

$$\sum_{x \in R \cup B} \sum_{y \in R \cup B} p_{xy} \leq \sum_{r \in R} g_r + \sum_{b \in B} g_b. \tag{3.27}$$

Our objective to maximize the energy efficiency is equivalent to minimizing the brown energy usage. Taking all the constraints discussed above, we eventually obtain an MILP as:

MILP :

$$\min : \sum_{r \in R} p_{er} + \sum_{b \in B} p_{eb},$$

$$\text{s.t.} : (3.21), \ (3.23), \ (3.24)-(3.27),$$

$$x_{rb} \in \{0, 1\}, \ y_b \in \{0, 1\}, \ \forall r \in R, b \in B.$$

3.3.3 A Two-Phase Heuristic Algorithm

It would be computationally prohibitive to solve the MILP problem for large-scale networks. To tackle this issue, we propose a two-phase polynomial-time heuristic algorithm, as summarized in Algorithm 2, in this section.

Phase I The first phase of our algorithm is to find out a green energy generation aware RRH-BBU association relationship, intending to minimize the number of activated BBUs and maximize the usage of local green energy. By minimizing the number of activated BBUs, the overall energy consumption could be lowered, potentially lowering the brown energy usage. Maximizing the local green energy usage is to lower the energy loss due to inter-DER energy transferring. Based on such principle, we design the first phase as shown in lines 1–13. In order to maximize the usage of local green energy, we try to assign BBUs from the one with the highest green energy generation rate. Therefore, we first sort the BBUs into set $sortedBBUs$ according to their green energy generation rates $g_b, \forall b \in B$ in a descend order in line 2. Thereafter, we try to sequentially assign the RRHs to BBUs until all RRHs are completely assigned in lines 3–13. For each RRH, we assign it to the first BBU with enough processing capacity to satisfy its demand in the sorted BBU set $sortedBBUs$. For each assignment, we correspondingly set the values of association relationship x_{rb} (line 6) and activation status y_b (lines 7–9), respectively. Then, we update the accumulated workload and the remaining processing capacity on the chosen BBU in line 10.

Phase II In phase II, we take the green energy generation aware RRH-BBU association relationship $x_{rb}, \forall r \in R, b \in B$ and BBU activation decisions $y_b, b \in B$ as input to plan energy transferring among the DERs. As the association relationship (i.e., the values of $x_{rb}, \forall r \in R, b \in B$) has already been known, it is possible to calculate the energy surplus on each network entity. The energy surplus a_b and b_r on BBU $b \in B$ and RRH $r \in R$ can be calculated as:

Algorithm 2 Two-phase heuristic RRH-BBU association and energy sharing scheme

---

Require: B, g_b, $\forall b \in B$, R, λ_r, g_r, $\forall r \in R$, η_{xy}, $\forall x, y \in B \cup R$
Ensure: x_{rb}, y_b, $\forall r \in R$, $b \in B$, p_{xy}, $x, y \in B \cup R$
1: Phase-I:
2: Sort BBUs B decreasingly according the green energy generation rates g_b, $\forall b \in B$ into set $sortedBBUs$
3: **for all** $r \in R$ **do**
4:     **for all** $b \in sortedBBUs$ **do**
5:         **if** BBU b has enough process capacity to serve RRH r **then**
6:             $x_{rb} \leftarrow 1$
7:             **if** $y_b == 0$ **then**
8:                 $y_b \leftarrow 1$
9:             **end if**
10:             Update the accumulated workload μ_b and the remaining process capacity on b, according to the request from r
11:             break
12:         **end if**
13:     **end for**
14:
15:     Phase-II:
16:     Calculate the surplus energy, i.e., a_r, a_b, $\forall r \in R$, $b \in B$ for all network entities
17:     Solve the LP

$$\min : \sum_{r \in R} p_{er} + \sum_{b \in B} p_{eb},$$

$$\text{s.t.} : (c1), (c2), (c3)$$

to obtain the energy transferring decisions p_{xy}, $x, y \in R \cup B$
18: **end for**

$$a_b = g_b - \alpha \sum_{r \in R} x_{rb} \lambda_r, \tag{3.28}$$

and

$$a_r = g_r - \beta \lambda_r, \tag{3.29}$$

respectively. Note that the energy surplus could be negative if the green energy generation rate fails to catch up with the energy demand. Actually, the negative energy surplus refers to the additional energy demand, which can be satisfied by either the brown energy or green energy from the other DERs with positive energy surplus.

Then, based on the energy surplus information and the energy transferring loss ratio, we can establish a linear programming (LP) model to maximize the usage of the surplus energy for overall energy efficiency as follows:

LP :

$$\min : \sum_{r \in R} p_{er} + \sum_{b \in B} p_{eb},$$

$$\text{s.t.} : \sum_{r \in R} p_{rb}\eta_{rb} + \sum_{x \in B} p_{xb}\eta_{xb} + p_{eb} \geq a_b, \forall b \in B, \quad (c1)$$

$$\sum_{x \in R} p_{xr}\eta_{xr} + \sum_{b \in B} p_{br}\eta_{br} + p_{er} \geq a_r, \forall r \in R, \quad (c2)$$

$$\sum_{x \in R \cup B} \sum_{y \in R \cup B} p_{xy} \leq \sum_{r \in R} a_r + \sum_{b \in B} a_b, \quad (c3)$$

where constraints (c1) and (c2) are to ensure all the additional energy demands get satisfied, and constraint (c3) is to ensure the green energy to be transferred does not exceed to the total available one. The objective is set the same as in Sect. 3.3.2. Accordingly, we solve the LP in line 17 to obtain the optimal energy sharing for the association results obtained in Phase I.

3.3.4 Performance Evaluation Results

In this section, we present our performance evaluation results on our proposed algorithm ("Two-Phase") by comparing it with the optimal solution ("Optimal") and other two competitors to verify the correctness of our joint optimization design and the efficiency of our proposal. The first competitor ("RRH-BBU Opti") minimizes the number of activated BBUs and fully explores the green energy generated at each DER, but without energy sharing among DERs. While, the second one ("Energy Sharing") emphasizes the energy sharing but does not optimize the RRH-BBU association. The simulated network environment strictly follows the system model presented in Sect. 3.3.1. In order to extensively evaluate the efficiency of our proposal, the request rates on RRHs, the green energy generation rates on RRHs and BBUs are all randomly generated with different seeds in each simulation. The basic network settings are as follows: $|R| = 40$, $|B| = 15$, $e_b = 1.0$, $e_r = 0.1$, $\lambda_r = 1.0$, $\bar{g}_b = 2.0$, $\tilde{g}_r = 1.0$, $\alpha = 4$, $\beta = 2$, where $|\cdot|$ and $\bar{\cdot}$ are the cardinality function and average function, respectively. Several groups of experiments have been conducted to investigate the effect of different network parameters on the energy efficiency. Each group of experiment includes 50 simulation instances, whose average brown energy consumption and total energy consumption are calculated.

Let us first check the effect of the static energy consumption of BBU to the overall energy efficiency. Figure 3.5 shows the average brown energy consumption under different values of static energy consumption from 0.5 to 5.0. We, respectively, plot the brown energy consumption and the total one in Fig. 3.5a, b, respectively, from which we can see that both are increased with the increasing of

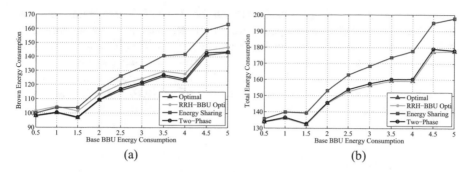

Fig. 3.5 The average energy consumption under different values of static BBU energy consumption. (**a**) Brown energy consumption. (**b**) Total energy consumption

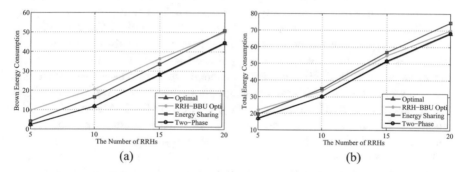

Fig. 3.6 The average energy consumption under different numbers of RRHs. (**a**) Brown energy consumption. (**b**) Total energy consumption

base energy consumption. Under all cases, we can see that our algorithm exhibits the highest energy efficiency compared with the other two competitors and performs much close to the optimal one. The advantage is even obvious when the base BBU energy consumption e_b is large. When e_b is small, minimizing the number of activated BBUs does not take much benefit to the overall energy efficiency and therefore "Energy Sharing" is even better than "RRH-BBU Opti" when $e_b \leq 0.5$. However, when e_b becomes dominant, it is significant to minimize the number of activated BBUs and to maximize the usage of each activated one and therefore "RRH-BBU Opti" becomes advantageous over "Energy Sharing." Nevertheless, our "Two-Phase" algorithm, thanks to the joint optimization of both factors, always performs the best under any value of e_b.

We then investigate the number of RRHs to the network energy efficiency by varying the number of RRHs from 5 to 20. The performance evaluation results are shown in Fig. 3.6, where the brown energy consumption and the total one are plotted in Fig. 3.6a, b, respectively. We can see that both the brown energy consumption and the total energy consumption increase with the number of RRHs. This is because more RRHs imply higher communication request demand and hence higher workload and energy consumption on the BBUs. It is interesting to notice that "Energy Sharing" is advantageous over "RRH-BBU Opti" on the brown energy but disadvantage on the overall energy when $|R| \geq 10$. This is because when the number of RRHs is large, carefully activating BBUs is beneficial to minimizing the energy consumption but it is also significant to share the green energy to reduce the brown energy consumption at the same time. More RRHs imply higher surplus energy at RRHs and therefore less brown energy will be used.

3.4   Conclusion

CRAN has already shown promising potential in both the performance efficiency and energy efficiency in the next generation wireless networks. By adopting renewable green energy to power the lightweight RRHs and BBUs, we think that it can further promote the network energy efficiency. With respect to the green energy generation rate diversity, it is significant to carefully activate the RRHs and manage the BBU-RRII association to maximize the green energy usage and hence minimize the brown energy consumption. In this chapter, we first investigate how to activate the RRHs in a green energy aware way for CoMP in CRAN. The problem is formulated into a non-convex optimization problem. We also propose a heuristic algorithm based on an ordered selection method. Then, we propose a BBU management algorithm that can minimize the BBU energy consumption, with a special emphasis on the joint optimization of RRH-BBU association and green energy sharing.

Chapter 4
Software Defined Networking I: SDN

Abstract Software defined network (SDN) is a newly emerging network architecture with the core concept of separating the control plane and data plane. Centralized controller is introduced to manage and configure network equipments to realize flexible control of network traffic and provide a good platform for application-oriented network innovation. It thus be able to improve network resource utilization, simplify network management, reduce operating cost, and promote innovation and evolution. Routing is always a major concern in network management. When SDN devices (e.g., OpenFlow switches) are introduced, routing becomes different. The flow table is usually implemented in expensive and power-hungry Ternary Content Addressable Memory (TCAM), which is thus capacity-limited. How to optimize the network performance in the consideration of limited TCAM capacity is therefore significant. For example, multi-path routing (MPR) has been widely regarded as a promising method to promote the network performance. But MPR is at the expense of additional forwarding rule, imposing burden on the limited flow table. On the other hand, a logical centralized programmable controller manages the whole SDN by installing rules onto switches. It is widely regarded that one controller is restricted on both performance and scalability. To address these limitations, pioneering researchers advocate deploying multiple controllers in SDNs where each controller is in charge of a set of switches. This raises the switch-controller association problem on a switch shall be managed by which controller. In this chapter, we first investigate how to schedule MPR with joint consideration of forwarding rule placement. Then, we study a minimum cost switch-controller association (MC-SCA) problem on how to minimize the number of controllers needed in an SDN while guaranteeing the flow setup time.

4.1 Bckground on Software Defined Networks

In traditional network architecture, once the network is deployed, reconfiguration network equipments (e.g., routers, switches, firewall, etc.) to satisfy the changing communication demand is a very complicated task. Now, high stability and high performance of network is not enough to meet the business requirements any more, flexibility and agility are even more crucial. We urgently need a new network architecture innovation to catch up with the increasing demand. Software defined networking (SDN) has emerged as a new paradigm that separates the data plane (forwarding function) from control plane. By such means, the whole network can be administered by centralized controller using a uniform programming interface as it shields the differences from the underlying network devices. The network thus becomes more flexible and intelligent, making it quite easy to deploy new network protocols [33]. The control plane is completely open. Users can freely customize their network routing strategy according to the application requirements. SDN fills the gap between the application and the network. It is without-doubt the development trend of the future network.

4.2 Forwarding Rule Management in SDN

To computer networks, routing is always major concern and has attracted much attention from scientists and engineers. With the recent prosperousness of computer networks, pioneering researchers found that single-path routing fails to explore the increasing capacity of communication networks. Multi-path routing (MPR), which can find many available paths between communication pairs, was proposed as a promising solution. It has already been proved that MPR is more efficient in improving the utilization of network bandwidth, reducing obstruction and achieving load balancing. Multi-path routing therefore has attracted many interests in the literature.

It is therefore natural to introduce MPR into SDN for network performance promotion. However, some new challenges are introduced. In SDN-enabled devices (e.g., OpenFlow switches [77]), the flow tables are made by expensive and power-hungry Ternary Content Aware Memory (TCAM), which is size-limited [20, 28]. This limits the number of forwarding rules that can be placed in the flow table. Note that MPR requires forwarding rule on all the routers involved. Although selecting more paths implies high network performance, it also requires more forwarding rules. As a result, in SDN, MPR shall be carefully scheduled with the consideration of size-limited flow table. Actually, many efforts have been devoted to addressing the limited TCAM size. Most existing studies on forwarding rule optimization, e.g., forwarding rule placement, mainly focus on single-path routing. Taking MPR into consideration, how to jointly schedule the routing and forwarding rule placement is still under-investigated.

In this section, we are motivated to study the MPR scheduling jointly with the forwarding rule placement in size-limited flow tables. Specially, we are interested in maximize the communication satisfaction with the consideration of communication demands. We first build an integer linear programming (ILP) model to describe the optimization problem, with joint consideration of MPR scheduling and forwarding rule placement. To address the computation complexity of solving ILP, we further propose a three-phase heuristic algorithm in polynomial time.

4.2.1 System Model and Problem Statement

In this work, we consider a network that can be described as a graph $G = (V, E)$, where V is a set of nodes and E denotes the links between these nodes. Figure 4.1 gives an illustrative network example with 9 nodes and 13 links. Among these nodes, there is a set D of communication pairs. We denote a pair using its source and destination $(s, d) \in D$ where $s, d \in V, s \neq d$. Each pair (s, d) is associated with a traffic demand Π^{sd}. A traffic flow originating from s may go through many intermediate relay node (i.e., routers) until reaching destination d. We suppose all the relay nodes are SDN-enabled (e.g., OpenFlow switches). To customize the routing strategy, forwarding rules must be placed at the corresponding routers according to the routing requirement. In MPR, the nodes on all forwarding paths for all sub-flows shall be placed with corresponding forwarding rules. Due to the capacity limitation of TCAM, the number of forwarding rules that can be placed on a node is limited. We denote the number of forwarding rules that can be placed on node $v \in V$ as S_v. Besides, the communication links are also capacity-limited. The capacity of link $(u, v) \in E$ is denoted as C_{uv}. For example, we have $C_{01} = 10$ in Fig. 4.1.

To the requirements of the communication pairs, we are interested in a max-min fairness scheduling which leads to high satisfaction. We define the satisfaction of a pair as the ratio between the achieved throughput and the desired demand. When the provisioned resources are not enough to satisfy the total communication demands, high satisfaction ensures high throughput with max-min fairness. On the other hand, when the achieved throughput is beyond the desired demand, high satisfaction indicates high residual capacity and hence high demand jitter tolerance. Therefore, it is significant to investigate how to maximize the minimum satisfaction of all communication pairs with joint consideration of MPR and forwarding rule placement in size-limited flow tables.

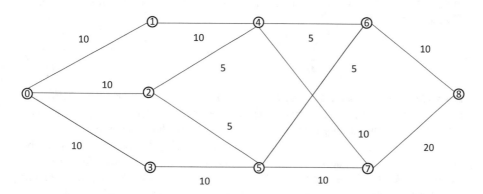

Fig. 4.1 A network example with 9 nodes and 11 links

4.2.2   Problem Formulation

Based on the system model as described above, we next present an integer linear programming (ILP) model to describe the problem studied in this paper.

Flow Conservation Constraints   As we have known, the traffic flow from node s may go across many different paths consisting of different links to its wanted destination d. We therefore define the flow of communication pair (s, d) on link (u, v) as f_{uv}^{sd}. According to the network flow theory, flow conservation must be reserved at all intermediate relay nodes, except the source and the destination nodes. As a result, we shall have

$$\sum_{(u,v)\in E} f_{uv}^{sd} = \sum_{(v,w)\in E} f_{vw}^{sd}, \forall (s, d) \in D, v \in V. \tag{4.1}$$

Note that the above flow conservation constraints cannot be applied to the source and destination nodes, to which we have the following constrains:

$$\sum_{v\in V} f_{sv}^{sd} = F^{sd}, \forall (s, d) \in D, \tag{4.2}$$

and

$$\sum_{v\in V} f_{vd}^{sd} = F^{sd}, \forall (s, d) \in D, \tag{4.3}$$

respectively, where float variable F^{sd} indicate the throughput that can be achieved.

Link Capacity Constraints   A traffic flow may go through many different links. While, a link may have sub-flows of many different communication pairs. Nevertheless, the overall flow amount through a link shall not exceed its link capacity. That is,

$$\sum_{(s,d)\in D} f_{uv}^{sd} \leq C_{uv}, \forall (u, v) \in E. \tag{4.4}$$

Flow Table Size Constraints   As we have known, besides link capacity constraints, SDN additionally introduces new constraints on the size of flow tables, indicating the number of forwarding rules that can be stored on the node. Whenever a flow or a sub-flow injects into an SDN node, a corresponding forwarding rule must be placed onto it. Note that no matter how many outgoing sub-flows are scheduled, we always need only one forwarding rule. Let binary variable x_v^{sd} denote whether a forwarding rule of pair $(s, d) \in D$ must be placed on node v. We shall have the following relationship:

$$x_v^{sd} = \begin{cases} 1, \text{ if } \exists f_{uv}^{sd} > 0, u \in V, \\ 0, \text{ otherwise.} \end{cases} \forall (s, d) \in D, v \in V. \qquad (4.5)$$

The above relationship can be translated into the following equivalent linear expressions as:

$$\frac{\sum_{(u,v)\in E} f_{uv}^{sd}}{A} \leq x_v^{sd} \leq \sum_{(u,v)\in E} f_{uv}^{sd} \cdot A, \forall (s, d) \in D, v \in V, \qquad (4.6)$$

where A is an arbitrary large number.

For each node v, the total number of rules that shall be placed onto it shall not exceed its flow table size S_v, i.e.,

$$\sum_{(s,d)\in D} x_v^{sd} \leq S_v, \forall v \in V. \qquad (4.7)$$

ILP Formulation Remember that our objective is to maximize the minimum satisfaction of all communication pairs. It is straightforward to define the satisfaction of a communication pair $(s, d) \in D$ as $\frac{F^{sd}}{D^{sd}}$. Taking all the pairs into consideration, the objective can be described as:

$$\max \min : \frac{F^{sd}}{D^{sd}}, \forall (s, d) \in D. \qquad (4.8)$$

In order to simplify the max-min objective into linear form, we define an auxiliary parameter Q with the following constraints

$$Q \leq \frac{F^{sd}}{D^{sd}}, \forall (s, d) \in D. \qquad (4.9)$$

Thus, by taking all the constraints discussed above, we obtain an ILP formulation of the max-min fairness problem as

$$\max : Q$$
$$\text{s.t. } (4.1)\text{–}(4.4), (4.6), (4.7), (4.9). \qquad (4.10)$$

4.2.3   A Three-phase Heuristic Algorithm

It is computationally prohibitive to solve the optimal solution of the ILP problem formulated in last section for large-size SDNs. To tackle this issue, we propose a three-phase polynomial-time heuristic algorithm in this section. In the first phase,

---

Algorithm 3 Find all potential paths

Require:
  $G = (V, E), C_{uv}, \forall(u, v) \in E, D.$
Ensure:
  path set *Allpath* for every (s, d)
 1: Initially, $Allpath = \emptyset$
 2: Exclude constraints (6) and (7) to obtain a LP problem
 3: Solve the LP to obtain $f_{uv}^{sd}, \forall(s, d) \in D, (u, v) \in E$
 4: **for all** $(s, d) \in D$ **do**
 5: **for all** $(u, v) \in E$ **do**
 6:   **if** $f_{uv}^{sd} > 0$ **then**
 7:    save the f_{uv}^{sd} and link (u, v) to *Allpath* as a new path ;
 8:    **for** each path in *Allpath*$[s, d]$ **do**
 9:     sort all nodes in *Allpath*$[s, d]$;
10:    **end for**
11:   **end if**
12: **end for**
13: **end for**

---

by excluding the TCAM size constraints, we first derive the MPR scheduling that can maximize the minimum satisfaction when the flow table is unlimited. We then further take the TCAM size constraints into consideration by checking whether all the nodes satisfy the capacity limitations. Only when no node in a path violates the flow table size constraint, it is feasible; otherwise, it is infeasible and needs to be rescheduled. To adjust these infeasible paths into feasible one, we discard some sub-flows to make sure that no TCAM size limitation is violated. Further, some sub-flows are merged into the switches with residual flow table capacity to maximize the user satisfaction. The three phases are summarized in Algorithms 1, 2, and 3, respectively.

Phase 1 (Algorithm 3): We first exclude the flow table size constraints by excluding (4.6) and (4.7) in ILP. As the integers are only involved in (4.6) and (4.7). A linear programming (LP) problem is thus obtained. By solving the LP problem, we obtain an MPR scheduling without flow table size constrains. By checking the values of $f_{uv}^{sd}, \forall(s, d) \in D, (u, v) \in E$, we can get whether a link, e.g., $(u, v) \in E$, is involved in the routing of flow, e.g., $(s, d) \in D$, or not. We save the links involved in the routing scheduling (lines 5–7). After that, we sort all node in corresponding path to get a full path from line 7 to 8.

Phase 2 (Algorithm 4): In this phase, we try to test whether the nodes in the path involved in MPR, according to the results of Algorithm 1, satisfies the flow table size constraint. As shown in Algorithm 2, we denote the residual flow table size as $Temp_S$, which is initialized with total flow table size S (line 1). By checking whether all the nodes involved in a path are with residual flow table size $Temp_S$ larger than 0, we can judge whether the path is feasible or not. If it is feasible, we subtract the residual flow table size of all the involved nodes by one to indicate that one forwarding rule is occupied. Otherwise, the path is marked as infeasible and shall be rescheduled.

Algorithm 4 Test the feasibility of all paths

Require:
   all paths $Allpath$ for every (s, d) ;
   flow $f_{uv}^{sd}, \forall (s, d) \in D, \forall (u, v) \in E$;
   flow table size $S_v, \forall v \in V$;
Ensure:
   dissatisfied paths $dispath$ for every (s, d)
   residual flow table capacity $Temp_S$
1: $Temp_S = S$
2: **for all** $(s, d) \in D$ **do**
3:    **for all** path p^{sd} in $Allpath$ **do**
4:       **if** all nodes' rule capacity in this path > 0 **then**
5:          the path p^{sd} is feasible ;
6:          **for** each node $u \in p^{sd}$ **do**
7:             $Temp_S[u] = Temp_S[u] - 1$
8:          **end for**
9:       **else**
10:          the path p^{sd} is infeasible;
11:       **end if**
12:    **end for**
13: **end for**

Algorithm 5 Reschedule and update the throughput

Require:
   all paths $Allpath$ for every (s, d) ;
   infeasible paths $dispath$ for every (s, d) ;
   flow f_{uv}^{sd} $\forall (s, d) \in D, \forall (u, v) \in E$;
   Residual flow table size $Temp_S$;
Ensure:
   MPR scheduling and rule placement results
1: **for all** path $p^{sd} \in dispath$ **do**
2:    **if** there is a feasible path candidate p'^{sd} in $Allpath$ **then**
3:       migrate the traffic of path p^{sd} to p'^{sd}
4:       update the throughput
5:    **else**
6:       remove the path p^{sd}
7:    **end if**
8: **end for**

Phase 3 (Algorithm 5): In this phase, we reschedule all the infeasible paths found out in Phase 2, by migrating their forwarding task to another feasible path in the set found in Phase 1. Similarly, we check both the residual link capacity and flow table size constraints to verify the feasibility of the potential path candidates. If no constraint is violated, we view the path as feasible (line 3) and migrate the traffic task to the feasible path (line 4). At the same time, the infeasible path is removed from the MPR scheduling (line 7). Otherwise, if no path candidate is found, we have to discard the infeasible path.

4.2.4 Performance Evaluation

To evaluate the performance of the proposed rule placement method, we have a test simulation. Usually, in the area of the simulation, there should be more than 10,000 nodes average. To simplify the simulation, we use a smaller size of nodes by scaling down other parameters like the rule capacity of each switch. We set the number of nodes is 200 and link capacity from 500 to 2000, while the average capacity of each link is about 10 K in SDN network [103]. The simulation proved that compared to traditional multi-stream table schema, the throughput forwarding to SDN network has been significantly improved.

According to the algorithm variables, we can see that objective minimal throughput is related to capacity of flow table TCAM size, link capacity, and size of the network topology. Therefore, we run a lot of experiments to explore the relationship between the target and the above-mentioned variables. Furthermore, the optimal results "Optimal" are obtained by solving the ILP problem using commercial solver Gurobi optimizer.

4.2.4.1 On the Effect of TCAM size

In the case of link capacity and network size, nodenumber = 2000 is a certain value, consider the impact of the flow table capacity TCAM size to minimal throughput.

As can be seen from the results in Fig. 4.2a–c, regardless of the size of the link capacity, the average maximum throughput of the network is increased firstly and finally reached a steady. For each curve, the reason for this trend is that, when the TCAM size increases, since the beginning of value is relatively small, for all paths, the number of infeasible path found by the Algorithm 1 will be large. After carrying out the diversion or discard feasible path, objective value will be decreased, so with increasing of flow table size in earlier stage, the objective also increases; when the flow table size increases to a nearly certain TCAM size, the increased number of viable path will gradually become decrease, until near 0, so the target minimal throughput also will stabilize at a fixed value.

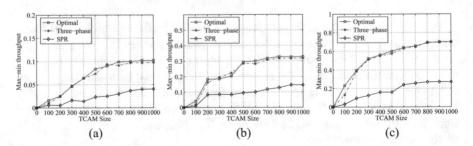

Fig. 4.2 Min-throughput under various values of TCAM size when $|N| = 200$. (**a**) link capacity = 500. (**b**) link capacity = 1000. (**c**) link capacity = 2000

We also explore the effect of TCAM size under .the link capacity is 500 M, 1000 M, and 2000 M. From another perspective, since a certain network topology, each node of the source to the demand for certain purposes, after flow table size increases to a certain value, the throughput of the whole network has reached the maximum, and therefore, according to these, we can choose the appropriate TCAM values for the topology of the network as well as other basic parameters. It will be the basis for subsequent experiments.

The figure shows that the result of three-phase algorithms and optimization models approximate are coincident, however for shortest-path routing(SPR), results may be only half of the before. TCAM size can determine the appropriate value based on the experimental results, which indicate that the proposed heuristic algorithm theory is feasible, and the result is ideal.

4.2.4.2 On the Effect of Link Capacity

Under certain conditions of TCAM size and network size (node number $= 2000$), we consider the impact of the link capacity to objective minimal throughput: As can be seen from the results in Fig. 4.3a–c, regardless of the size of the flow table size, the average maximum throughput of the network is increased. When the link capacity increase, on the whole, traffic on each path increase, which explains the reason why the lager of link capacity, the larger of maximum throughput(on condition that the demand is large enough). Analyzing each figure, we can see the result of three-phase algorithms is better than SPR. Therefore, in the actual network environment, the link capacity is also a key factor to improve network performance.

4.2.4.3 On the Effect of Network Size

The above two experimental network topology is fixed, when the node size becomes larger, the influence of network size changing to an objective cannot be ignored.

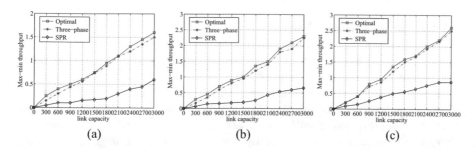

Fig. 4.3 Min-throughput under various values of link capacity. (**a**) $|N| = 200$, TCAM size $= 400$. (**b**) $|N| = 500$, TCAM size $= 600$. (**c**) $|N| = 500$, TCAM size $= 800$

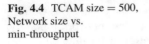

Fig. 4.4 TCAM size = 500,
Network size vs.
min-throughput

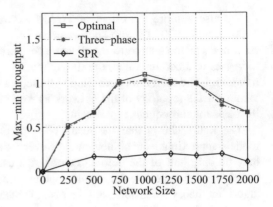

In the overall trend, with the increasing of flow table size, the average maximum throughput also increased, when the value of TCAM size increases to a certain stage, the target upward trend becomes slow, namely saturation. Therefore, when the network topology size is medium, TCAM size is 500–700, the average maximum throughput of the network can generally be maximized. As link capacity increasing, minimal throughput rises perpendicularly, thus the larger of link capacity, the larger of the target value.

Figure 4.4 shows the curve rise firstly, then a decline. The reason for this phenomenon is that the nodes are relatively small in earlier stage, and the preset flow table size is relatively large, which the network increasing has a big impact on the maximum throughput of the network, so as to increase the number of nodes, the target value also increased. When the network node reached about 750–1500, the results can be analyzed based on Experiment 1, relative to the current TCAM size, the most appropriate network size, the average throughput of whole network is close to the maximum, So curve is relatively flat during this time, after the number of network nodes more than 1500, the number of paths found by the algorithm will increase, but the capacity of the flow table is limited, lead to the increasing of number of infeasible paths. So with increasing of the number of nodes, TCAM size impact on target increases, resulting in a downward trend.

4.2.4.4   On the Effect of Demand

In the end, we studied the impact of the initial value *Demand* to the objective of maximizing the minimum throughput.

Figure 4.5 shows the downward trend in the curve, the reason is that demand *Demand* is only used to calculate the objective value, it is independent of the network topology, link capacity size, and flow table capacity size. Therefore, demand *Demand* only has a quantity relationship with the target, for nothing meaning to the target.

Fig. 4.5 Demand vs.
min-throughput

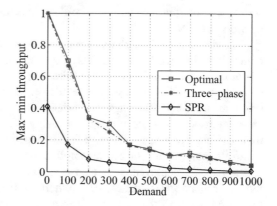

The model testing results are in good agreement with the algorithm described above, the shortest-path routing testing results are probably the half of the front. The conclusion is that the algorithm has a better performance. Therefore, it proved that the algorithm is feasible. Focusing on route optimization in SDN system, we make a deep study for OpenFlow switch architecture aiming at optimizing data transmission performance of OpenFlow switch. Experimental research focuses on optimization of multi-path routing and improvement of system performance on SDN. Summed it up, the idea of technical solutions is to take advantage of the flexibility of SDN, based on TCAM capacity restrictions, use multi-path routing, maximize network throughput, to improve system performance objectives.

4.3   Controller Management in SDN

In SDNs, the controller is responsible for determining the forwarding rules on the forwarding devices through a control channel. Therefore, the controller is often regarded as the central brain and plays an important role in SDNs. The most common SDN implementation adopts a centralized network control, where a controller manages and operates the network from a global view. Whenever a switch receives a new flow and finds no matching entry in the flow table, it immediately requires the controller to install appropriate forwarding rules along the desired flow path. However, in a large-scale SDN deployment, this rudimentary centralized approach has several limitations on both performance and scalability. On one hand, a single controller usually has a limited capacity and hence cannot handle large number of flows originating from all the infrastructure switches. In this case, some request packets may have to be dropped, incurring negative effect to the network performance. On the other hand, the latencies between the single controller and the switches situated at geographically distributed locations are highly varied. To some switches far away from the controller, long flow setup time may be introduced. This severely limits the network performance, especially to SDN-based wireless area networks (WANs).

To address these limitations, pioneering researchers advocate deploying multiple controllers that work cooperatively to manage network traffic flows [39, 109]. Much effort has been devoted to addressing various problems related to multiple controllers in SDNs. For example, Heller et al. [42] study the controller placement problem and analyze the impact of the controller locations on the average and worst-case controller-to-switch propagation delay. The controller placement problem is then extensively studied from different aspects, e.g., [45, 53, 78, 94, 110]. The controller placement solutions are applied in pre-deployment period. When multiple controllers are deployed, it is also essential to consider the association between the controllers and switches, i.e., switch-controller association, as it also affects the performance of the network. Specially, to make the network cost-efficient, one feasible solution is to minimize the number of activated controllers. Of course, no matter how many controllers are actually used, it is first required that the flow setup time is guaranteed. Therefore, we are motivated to investigate the flow setup time aware switch-controller association problem aiming at minimizing the number of controllers, i.e., minimum cost switch-controller association (MC-SCA) problem. In this section, we first formulate MC-SCA with the consideration of transmission time between switches and controllers into a quadratic integer programming (QIP) problem, which is then transformed into an equivalent integer linear programming (ILP) problem. We also formally prove that MC-SCA is NP-Hard and then propose a heuristic algorithm with high efficiency and low complexity.

4.3.1 System Model

We consider an SDN shown in Fig. 4.6. Following the OpenFlow model [66], the network consists of a controller plane and a data plane. The controller plane is formed by a set I of controllers. The data plane is composed of a set J of OpenFlow switches which forward data flow according to the flow table. Without loss of generality, we assume that a switch can be reached from any controller. The transmission latency between a controller $i \in I$ and a switch $j \in J$ is denoted as d_{ij}.

To further state our problem, we first elaborate the flow setup process as follows. In SDNs, each switch has a flow table where each entry corresponds to the operation rule (e.g., forwarding, discarding, packet header altering, etc.) for a flow. When a packet arrives at the OpenFlow switch, the switch extracts its header information and then matches it with the flow table entries. If the matching is successful, the switch executes the forwarding decision instantaneously. If the switch does not contain a matching rule, the packet is sent to the associated controller requesting for an action to execute. The controller will determine the rule to handle the packet and respond to the switch request with an action to perform on all packets of this flow. A controller $i \in I$ can only manage the switches that associate to it.

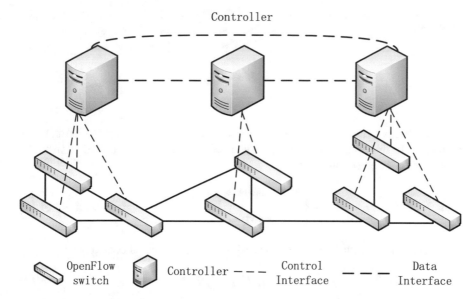

Fig. 4.6 Network model

From the perspective of controller, the flow setup requests may accumulate at the egress of controller and form a request queue. Following the SDN model presented in [52] and [121], we assume that a switch $j \in J$ generates flow setup requests following Poisson process with rate λ_j. The requests are processed by controller $i \in I$ with exponentially distributed service time with average value $1/\mu_i$, where μ_i is the average service rate.

4.3.2 QIP Formulation

To represent the association relationship, we first define a binary variable e_{ij} to denote whether switch j is associated with controller i as:

$$e_{ij} = \begin{cases} 1, \text{ if switch } j \text{ is associated with controller } i, \\ 0, \text{ otherwise.} \end{cases} \tag{4.11}$$

Completeness Constraints According to SDN philosophy, a switch must be associated with one controller such that it can be manipulated according to the communication requirements. That is,

$$\sum_{i \in I} e_{ij} = 1, \forall j \in J. \tag{4.12}$$

Fig. 4.7 Request queue at an
SDN controller

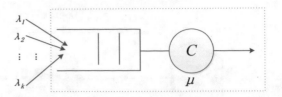

Note that a switch can be associated with a controller $i \in I$ provided that i is activated. To this end, we define a binary variable c_i to denote whether controller i is activated or not as follows:

$$c_i = \begin{cases} 1, \text{ if controller } i \text{ is activated,} \\ 0, \text{ otherwise.} \end{cases} \tag{4.13}$$

We must guarantee that each switch is assigned to an activated controller, i.e.,

$$e_{ij} \leq c_i, \forall i \in I, j \in J. \tag{4.14}$$

Flow Setup Time Constraints As we know, the first packet (i.e., flow setup request) of a flow arriving at a switch $j \in J$ but with no entry matching in the flow table must go through the associated controller of j, say $i \in I$, to obtain the rule for the flow. The flow setup time has a critical impact on the system performance as the flow can pass j only after the rule is ready. A controller takes charge of the flow setup requests from all the switches in its management domain. A flow setup request queue is formed at each controller as illustrated in Fig. 4.7. The request from any switch can be regarded as an individual and independent Poisson process. As the sum of a set of independent Poisson processes is still a Poisson process and the request handling time is exponentially distributed, we can describe the process of handling flow setup requests on a controller using $M/M/1$ queuing model. Therefore, the queuing time at a controller i can be calculated as:

$$\frac{1}{\mu_i - \sum_{k \in J} \lambda_k e_{ik}},$$

where $\sum_{k \in J} \lambda_k e_{ik}$ denotes the sum Poisson process arrival rate. To ensure that the queue is steady, a hidden condition must be met is that the sum of the arrival rate of all switches shall not be beyond the service rate provided by a controller. Therefore, we have

$$\mu_i - \sum_{k \in J} \lambda_k e_{ik} > 0. \tag{4.15}$$

From the perspective of a switch, the flow setup time shall also take the transmission latency between its associated controller. To ensure the network performance, the total flow setup time experienced by any switch shall not exceed T, i.e.,

$$e_{ij}d_{ij} + \frac{1}{\mu_i - \sum_{k \in J} \lambda_k e_{ik}} \leq T, \forall i \in I, j \in J. \tag{4.16}$$

When $\mu_i - \sum_{k \in J} \lambda_k e_{i,k} > 0$, (4.16) can be transformed into

$$\mu_i e_{ij} d_{ij} + \sum_{k \in J} T\lambda_k e_{ik} - d_{ij} \sum_{k \in J} \lambda_k e_{ij} e_{ik} \leq T\mu_i - 1, \tag{4.17}$$

$$\forall i \in I, j \in J.$$

QIP Formulation We intend to minimize the number of activated controllers, which can be expressed as:

$$\sum_{i \in I} c_i.$$

Summarizing the above together, we can formulate the MC-SCA problem as:

$$\min : \sum_{i \in I} c_i,$$

$$\text{s.t. : } (4.12), (4.14), (4.15), (4.17),$$

$$c_i \in \{0, 1\}, e_{ij} \in \{0, 1\}, \forall i \in I, j \in J,$$

which is a QIP as there are quadratic terms, e.g., $e_{ij}e_{ik}$.

Theorem 4.1 *The flow setup time aware minimum cost switch-controller association problem is NP-Hard.*

Proof Let us consider a special case of the MC-SCA problem by excluding the queuing time at each controller. In this case, whether a switch can be associated with a controller is only determined by the transmission time between them. For example, for a controller $j \in J$ and a switch $i \in I$, only when $d_{ij} \leq T$, i can be associated with j. We shall find out the number of activated controllers that are able to ensure the completeness constraints in (4.12) that any switch is associated to a controller. This is exactly a minimum set cover problem [17], which has been proved as NP-Hard. ∎

4.3.3   QIP to ILP Transformation

We notice that it is possible to linearize the quadratic terms $e_{ij}e_{ik}$ by introducing new auxiliary variables

$$x_{ijk} = e_{ij}e_{ik}, \forall i \in I, j, k \in J$$

which can be equivalently replaced by the following linear constraints:

$$x_{ijk} \leq e_{ij}, \forall i \in I, j, k \in J \tag{4.18}$$

$$x_{ijk} \leq e_{ik}, \forall i \in I, j, k \in J \tag{4.19}$$

$$x_{ij} \geq e_{ij} + e_{ik} - 1, \forall i \in I, j, k \in J. \tag{4.20}$$

The constraint (4.17) can be then rewritten in linear form as:

$$\mu_i e_{ij} d_{ij} + \sum_{k \in J} T\lambda_k e_{ik} - \sum_{k \in J} \lambda_k x_{ijk} d_{ij} \leq T\mu_i - 1,$$

$$\tag{4.21}$$

$$\forall i \in I, j \in J.$$

Thus, an ILP formulation for the MC-SCA problem can be obtained as

$$\min : \sum_{i \in I} c_i,$$

s.t. : (4.12), (4.14), (4.18)–(4.21),

$$c_i \in \{0, 1\}, e_{ij} \in \{0, 1\}, x_{ijk} \in \{0, 1\}, \forall i \in I, j, k \in J.$$

4.3.4   Heuristic Algorithm

It is still computationally prohibitive to solve the ILP problem to get the optimal solution in large-scale SDNs. To address this problem, we propose a heuristic algorithm named "Allocation-Merge Algorithm" in this section.

Allocation-Merge Algorithm assigns switches to controllers according to the transmission latency under the constraints. The overall algorithm is presented in Algorithm 6, which mainly consists of three phases, initial association (lines 1–17), reassociation (lines 18–20), and merging (line 21). In the initial allocation phase, we first greedily associate each switch to the controller with the smallest transmission latency, provided that the setup time constraints are not violated with the consideration of queueing time by checking constraints (4.15) and (4.16). Therefore, after the initial association, there are some switches remaining in un-associated state. We shall then try to associate each of them onto an appropriate controller. In order to minimize the number of activated controllers, we first try to associate an un-associated switch to an activated controller. If no activated controller can accommodate the switch without violating constraints (4.15) and (4.16), we activate an inactive controller and associate the switch to it. After the second phase, we ensure that all the switches are successfully associated to appropriate controllers with guaranteed flow setup time. However, we notice that the performance can be

---

Algorithm 6 Allocation-merge algorithm

Require: controller set I, switch set J, the transmission delay $d_{ij}, i \in I, j \in J$, the maximum allowable setup time T, flow request rate $\lambda_j, j \in J$, the processing rate $\mu_i, i \in I$

Ensure: the number of activated controllers

1: Initialize three arrays: $e[i, j], c[i], s[j], i \in I, j \in J$ as 0
2: Select the controller with the smallest transmission latency d_{ij} for each switch
3: $e[i, j] \leftarrow 1, c[i] \leftarrow 1, s[j] \leftarrow 1$
4: **for all** $i \in I$ **do**
5:     Calculate $m = \mu_i - \sum_{k \in J} \lambda_k e_{i,k}$
6:     **if** $m > 0$ **then**
7:         **for all** $j \in J$ **do**
8:             Calculate $t = e_{i,j} d_{i,j} + \frac{1}{m}$
9:             **if** $t > T$ **then**
10:                 $e[i, j] \leftarrow 0, s[j] \leftarrow 0$
11:             **end if**
12:         **end for**
13:     **else**
14:         Exclude the switches with the smallest arrival rates from i until $m > 0$
15:     **end if**
16: **end for**
17: $\forall i \in I$, if no switch assigned to i, $c[i] \leftarrow 0$.
18: **for all** $j \in J$ and $s[j] = 0$ **do**
19:     Associate j to an activated controller provided that (4.15) and (4.16) are satisfied.
20: **end for**
21: Merge the activated controllers provided that (4.15) and (4.16) are satisfied.

---

further improved as some controllers are with small request load after the initial two phases. Therefore, we then try to merge the activated controllers with light loads to reduce the number of the controller. We iteratively merge two controllers into one provided that the flow setup time constraints are not violated. Finally, we obtain the controller activation and switch-controller association decisions.

4.3.5 Performance Evaluation

In this section, we present our simulation-based performance evaluation results on the efficiency of our proposed algorithm, by comparing it (i.e., "AMA") against the optimal solution and another greedy-based heuristic algorithm "GA." The optimal results "Optimal" are obtained by solving the ILP using commercial solver Gurobi optimizer [36]. The basic idea of the greedy algorithm is to assign as many switches as possible to an activated controller without violating the resource capacity constraints and one controller is activated in each iteration. The whole process works as follows. At first, we decreasingly sort both the switches and the controllers according to their request rates and processing rates, respectively. We then iteratively activate the controllers and assign as many switches as possible to the activated controller in each iteration. At the same time, constraints (4.15)

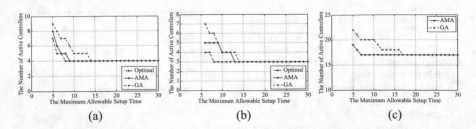

Fig. 4.8 On the effect of the maximum allowable flow setup time (**a**) |I| = 10, |J| = 20, λ = 8, μ = 50 (**b**) |I| = 10, |J| = 20, λ ∈ [5, 10], μ ∈ [40, 60] (**c**) |I| = 25, |J| = 100, λ = 8, μ = 50

and (4.16) are checked to see whether the setup time constraints are satisfied or not. The iterative process stops when all switches are assigned.

To extensively investigate the performance of our heuristic algorithms, we vary the values of the number of controllers $|I|$, the number of switches $|J|$, the arrival rate of the flow request λ (here we consider uniform request rate for all switches), the maximum allowable flow setup time T in different group of simulations. Twenty simulation instances in each group are conducted to get the average number of activated controllers. In each simulation instance, the transmission time between switches and controllers is randomly generated.

We first check how the maximum allowable setup time T affects the number of controllers. Figure 4.8 presents the simulation results. In this group of simulations, we fix $|I| = 10$, $|J| = 20$, λ = 8, μ = 50 and vary T from 1 to 30. The results are presented in Fig. 4.8a. In Fig. 4.8b, we set $|I| = 10$, $|J| = 20$, λ ∈ [5, 10], μ ∈ [40, 60] and T from 1 to 30. From both Fig. 4.8a, b, we can see that our proposed algorithm much approaches the optimal one and outperforms the greedy one, under any settings. Besides, it can be also noticed that the minimum cost shows as a decreasing function of the maximum allowable setup time. One more interesting thing is that all three algorithms converge and have the same number of controllers when the maximum allowable setup time is big enough. This is because, longer queuing time is tolerable with larger value of T. In this case, more requests can be allocated to one controller and hence less activated controllers are needed. In order to further study the performance of our algorithms, we extend the scale of the network under the setting of $|I| = 25$, $|J| = 100$, λ = 8, μ = 50 and T from 1 to 30. Due to the complexity of obtaining the optimal results, only "AMA" ad "GA" are reported in Fig. 4.8c. We can still see that "AMA" still performs better than "GA" in large-scale networks.

Then, we study the effect of the number of switches to the minimum number of the activated controllers using different settings. Similar to the study on the value of maximum allowable setup time, three group of settings are also considered. In the first group of simulations, we fix $|I| = 10$, λ = 8, μ = 50, $T = 8$ and vary the number of switches from 1 to 30 and show the results in Fig. 4.9a. In the second group of simulations, we further vary the values of the flow request arrival rate λ ∈ [5, 10], the processing rate of controller μ ∈ [40, 60] and show the results

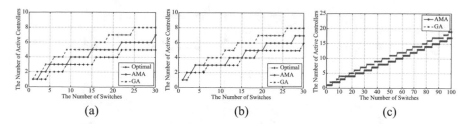

Fig. 4.9 On the effect of the number of switches

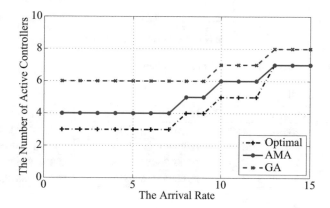

Fig. 4.10 On the effect of flow setup request arrival rate

in Fig. 4.9b. Figure 4.9c gives the results under the settings of $|I| = 25$, $\lambda = 8$, $\mu = 50$, $T = 10$ and $|J|$ from 1 to 100. Once again, we can see that our algorithm performs much close to the optimal one and has obvious advantage over the greedy one. We also notice that the number of activated controllers needed increase with the number of switches. This is attributed to the fact that more controllers must be activated to ensure the setup time if there are more switches.

Next, Fig. 4.10 gives the results on the effect of the arrival rate under the setting of $T = 8$, $|I| = 10$, $|J| = 20$, $\mu = 50$ and λ from 1 to 15. From the figure, we can still see the high efficiency of our proposed Allocation-Merge Algorithm. In addition, it also shows that the number of the activated controllers increases with the increasing of the arrival rate. It is straightforward to know that more requests imply that more controllers shall be activated to handle them.

Finally, in order to extensively show the efficiency of our proposed algorithm, we carry out a group of experiments by randomly setting the request arrival rates on all switches in [5, 10], the processing rates on the controllers in [40, 60], and maximum allowable setup time in [5, 15] in different simulation instances. We plot the cumulative distribution function (CDF) of the number of activated controllers for 200 instances in Fig. 4.11, from which we still see the high efficiency of our algorithm under any random settings. After extensively validating the efficiency of our algorithm, we further conduct a group of experiments to check the computation

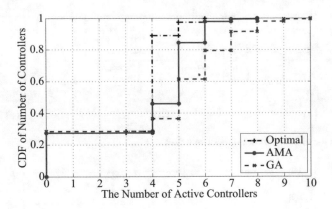

Fig. 4.11 CDF of the minimum number of controllers

Table 4.1 Running time

Instances	Optimal	Heuristic algorithm	Greedy algorithm
50	206.400718887	0.00139172755833	0.000393690880978
100	400.420835177	0.0026170858729	0.000931541952999
200	718.891972255	0.00520450670253	0.00142874264548
300	1279.7988438	0.00856356298393	0.00310085783613

time of different algorithms. The results are reported in Table 4.1. Obviously, both greedy and our heuristic algorithm require much less running time than solving the ILP to get the optimal solution. "AMA" needs only a little more running time than "Greedy."

4.4 Conclusion

In this chapter, we investigate two issues, i.e., rule management and controller management, in SDNs. We first focus on rule placement for multi-path routing in SDNs to select multiple feasible paths with the consideration of flow table size constraints. The objective is to maximize the minimum satisfaction of all flows. The problem is formulated into an ILP and a three-phase algorithm is proposed. We then investigate the MC-SCA problem on how to minimize the number of activated controllers with guaranteed flow setup time. By modeling the flow setup request process on a controller using an $M/M/1$ queue, we first formulate the MC-SCA into a QIP problem and further linearize it into an ILP problem. We also prove that the MC-SCA problem is NP-Hard. To address the computation complexity, a heuristic setup time aware and cost-efficient multiple controller placement algorithm is introduced in this chapter.

Chapter 5
Software Defined Networking II: NFV

Abstract Network function virtualization (NFV) emerges as a promising technology to increase the network flexibility, customizability, and efficiency by softwarizing traditional dedicated hardware based functions to virtualized network functions. The prosperous potential of edge cloud makes it an ideal platform to host the network functions. From the perspective of network service providers, an inevitable concern is how to reduce the overall cost for renting various resources from infrastructure providers. This first relates to the virtualized network function (VNF) placement, which shall not be discussed independently without the consideration of flow scheduling. In this chapter, we first discuss a static VNF placement problem with preknown service request rate. Then, we consider a more practical scenario and we alternatively investigate how to dynamically minimize the overall operational cost with joint consideration of packet scheduling, network function management, and resource allocation, without any prior knowledge.

5.1 Background on Network Function Virtualization

The concept of service chaining (SC) [54, 120] has been widely recognized as a promising way to provide flexible and cost-efficient network services using a sequential chain of network functions (NFs). For example, a typical firewall service may require an NF chain of a network address translation (NAT), a deep packet inspection (DPI), and an access control (AC). The newly emerged network function virtualization (NFV) technology softwarizes various NFs traditionally supported by dedicated hardware devices using lightweight virtualized network function (VNF) instances, which can be generally treated like virtual machines (VMs) and consolidated in general commercial-off-the-shelf (COTS) servers, e.g., ×86 servers. By such means, network services now can be realized by a service chain of VNF instances, which can be created, migrated, or killed anytime and anywhere, much promoting the network flexibility, customizability, and scalability. This naturally fits the requirements from big data processing.

5.2   Virtualized Network Function Placement

By exploring NFV technology, the capital expenditure (CAPEX) of service provider is much saved. Instead, the operational expenditure (OPEX) becomes dominant, especially in the consideration of the giant data volume in big data processing. How to lower the OPEX therefore naturally becomes one of the major concerns to network service providers. Network infrastructure providers usually charge different links with different unit costs, e.g., according to the link utilization. To reserve the network service semantics, the data generated from the source must go through a predetermined chain of VNF instances (e.g., DPI) before actual data processing. Consequently, VNF instance placement strategy imposes a deep impact on the OPEX as different placement strategies exhibit different communication patterns and hence the communication cost. Let us consider a network infrastructure with four nodes and two network flows with the flow rate of 1. Both flows go through an NF chain of DPI and AC. Suppose all the links are with unit cost 1. It can be easily seen from Fig. 5.1 that two possible placement strategies result in different communication costs of 10 and 8, respectively.

Although VM placement for communication cost minimization has been widely addressed in the literature, they can hardly be applied to VNF instance placement because they usually fail to capture the processing sequence of NF chains, i.e., network service semantics. To tackle this problem, several recent studies have been proposed. However, most existing work focuses on the minimization of deployment cost. For example, Moens et al. [73] investigate an NF placement to minimize the number of activated servers. Moreover, they do not take the flow balancing issue into consideration either. Previous work usually assumes that there exists only one VNF instance for each NF [1, 73], or the distributed flow rates are predetermined [68]. We argue that in big data processing, there shall exist thousands of hundreds of network flows characterized by large volume and geo-distribution. In this case, multiple VNF instances for the same NF shall be created and carefully placed to process the big data traffics efficiently. Furthermore, careful load balancing shall be conducted between these available NF instances.

In this section, we are motivated to investigate an efficient VNF instance placement problem to minimize the overall communication cost for big data processing. We first transform the multiple VNF instance placement problem into a VNF instance selection problem, which is then formally formulated into a mixed integer linear programming (MILP) problem. We also propose a relaxation-based low-complexity heuristic algorithm to address the computational complexity and validate the high efficiency of our algorithm through extensive experimental results.

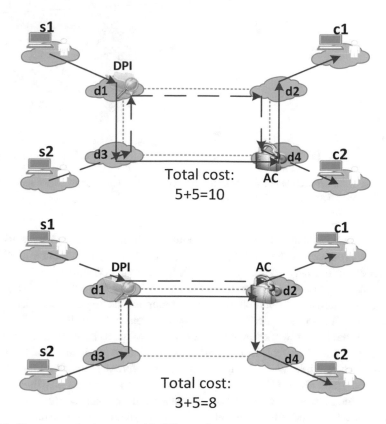

Fig. 5.1 The communication costs with different placement strategies

5.2.1   System Model

A network infrastructure can be represented by an indirected graph $G_d = (D, E_d)$, where D is a set of network nodes and E_d is a set of network edges. A node could be a server in a data center or one inside a base station. Each edge $e_{dp} \in E_d, d, p \in V_d$ is with a weight H_{dp} denoting communication unit cost between nodes d and p, which are predetermined. Specially, we set the communication unit cost between the same node as 0, i.e., $H_{dd} = 0, \forall d \in D$.

In big data processing, there are a large number of data flows originating from different geo-distributed sources and destined to different data processors, i.e., data consumers, for different analytics. We respectively denote the set of data flows, sources, and consumers as F, S, and C, respectively. During the data transferring, a data flow $f \in F$ with rate Λ_f shall go through a set N of predetermined network functions for different purposes, e.g., security enforcement. Taking all the flows and NFs into consideration, we can logically represent the data transferring process using a directed acyclic graph (DAG) $G_n = (V_n, E_n)$, where $V_n = S \cup N \cup C$.

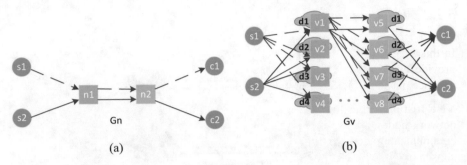

Fig. 5.2 Network service chain and extended network service chain. (**a**) Network service chain G_n. (**b**) Extended network service chain G_v

We define such DAG as network service chain graph in this paper. A network flow $f \in F$ is produced at the sources $s_f \in S$, processed by a sequence of NFs N with predetermined network service semantics, and finally consumed by a consumer $c_f \in C$. An edge $e_{nm} \in E_n$ represents the network flow from NF n to NF m. For example, the network service chain graph of our toy example in Fig. 5.1 can be represented as Fig. 5.2a. An NF is shared by different network flows. In Fig. 5.2a, we can see that two network flows from $s1$ to $c1$ and $s2$ to $c2$ both go through the same NF chain of $n1$ and $n2$. Unlike traditional dedicated hardware based NFs, one NF may have multiple VNF instances placed in different network nodes. We use an extended graph $G_v = (V_v, E_v)$ (e.g., Fig. 5.2a) to represent all possible instance placement and network flows based on the network flow graph G_n (e.g., Fig. 5.2b). The node set V_v includes three subsets sources S, extended VNF instances V, and destinations C, i.e., $V_v = S \cup V \cup C$. Different from the original network service chain graph, each NF n now has $|D|$ possible instances in different locations in the extended graph. As a result, totally there are $|N| \cdot |D|$ possible VNF instances in this network, i.e., $|V| = |N| \cdot |D|$. For each possible instance $v \in V$, let $d(v)$ and $n(v)$ denote its hosting node and NF type, respectively. For example, the hosting node of VNF instance $v6$ is $d2$ and NF type is $n2$, i.e., $d(v6) = d2$ and $n(v6) = n2$. Similarly, we denote $\bar{d}(d)$ and $\bar{n}(n)$ as the VNF set placed in network node d and the VNF set providing NF n, respectively, e.g., $\bar{d}(d2) = \{v2, v6\}$ and $\bar{n}(n2) = \{v5, v6, v7, v8\}$. We can see from Fig. 5.2b that we have totally $2 \cdot 4 = 8$ possible VNF instances located in 4 network nodes to provide 2 types of NFs. The network flow from NF n to m can be freely distributed among the VNF instances (u, v), as long as u and v provide NFs n and m, i.e., $u \in \bar{n}(n)$, $v \in \bar{n}(m)$. From the aspect of cost, we assume that for each NF n, totally A_n instances can be used. That is to say, we shall choose A_n VNF instances from the $|D|$ possible ones for each NF n such that the overall communication cost is minimized.

5.2.2 *Problem Formulation*

In this section, based on the proposed extended network service chain graph, we provide an MILP description of our problem with joint consideration of VNF instance placement and flow balancing.

VNF Instance Placement Constraints As each NF n has multiple instances in all network nodes, we are interested in how to choose from them to lower the total communication cost. Let binary variable x_v indicate whether a VNF instance v is chosen or not as

$$x_v = \begin{cases} 1, & \text{if possible instance } v \text{ is chosen,} \\ 0, & \text{otherwise.} \end{cases}$$

As mentioned in Sect. 5.2.1, the total number of chosen VNF instances for NF n is A_n. That is,

$$\sum_{v \in \bar{n}(n)} x_v = A_n, \forall n \in N. \tag{5.1}$$

By constraining the number of nodes that can be chosen from the extended graph G_v, it equivalently describes placing A_n instances for each NF in the network infrastructure G_d.

Flow Distribution Constraints In the extended network service chain graph G_v, the flow constraints of any VNF instance v can be represented by the relationships between the input flows from its parents $i(v)$ and output flows to its children $o(v)$. Let λ_{uv}^f denote the flow rate between any two VNF instances u, v for flow $f \in F$, if there exists an edge $e_{uv}, e_{uv} \in E_v$ in graph G_v.

If the rate of network flow processed by a VNF instance v is larger than 0, it indicates that this VNF instance to provide NF $n(v)$ is used and an instance must be placed in network node $d(v)$, $x_v = 1$. The relationship between x_v and the input flow rate f_{uv} can be therefore described as:

$$\frac{\sum_{f \in F} \sum_{u \in i(v)} \lambda_{uv}^f}{L} \leq x_v \leq \sum_{f \in F} \sum_{u \in i(v)} \lambda_{uv}^f \cdot L, \forall v \in V, f \in F, \tag{5.2}$$

where L is an arbitrary large number.

To guarantee that all network traffics are successfully delivered from the sources to the consumers, regardless of what kind of intermediate NFs are went through, we have

$$\sum_{v \in o(s_f)} \lambda_{s_f v}^f = \Lambda_f, \forall f \in F, \tag{5.3}$$

and

$$\sum_{u \in i(c_f)} \lambda_{uc_f}^f = \Lambda_f, \forall f \in F, \tag{5.4}$$

respectively.

Besides, for each intermediate VNF, the input flows from its input (or parent) nodes and its output (or child) nodes shall follow the relationship

$$\sum_{u \in i(v)} \lambda_{uv}^f = \sum_{w \in o(v)} \lambda_{vw}^f, \forall v \in V, f \in F. \tag{5.5}$$

to ensure the flow conservation.

A Joint MILP Formulation We are interested in minimizing the total communication cost between all pairs of connected VNF instances (u, v) as:

$$Cost_{com} = \sum_{f \in F} \sum_{e_{uv} \in E_v} \lambda_{uv}^f \cdot H_{d(u)d(v)}. \tag{5.6}$$

By summing up all above, we get the following **Com-Min** problem:

Com-Min :

$$\min : \sum_{f \in F} \sum_{e_{uv} \in E_v} \lambda_{uv}^f \cdot H_{d(u)d(v)},$$

$$\text{s.t. : } (5.1), (5.2), (5.3), (5.4), \text{ and } (5.5),$$

$$x_v \in \{0, 1\}.$$

Note that it is computationally prohibitive to solve this MILP problem due to the involvement of integer variables x_v, especially in large-scale network cases. To tackle this problem, we design a low-complexity heuristic algorithm in the next section.

5.2.3 Algorithm Design

It can be observed that the only binary variables involved in our **Com-Min** problem are x_v.

Algorithm 7 Relaxation-based algorithm

1: Relax the integer variable x_v, solve the **Com-Min-LP** problem

$$\textbf{Com-Min-LP} :$$

$$\min : \sum_{e_{uv} \in E_v} \lambda_{uv} \cdot H_{d(u)d(v)},$$

$$\text{s.t.} : (5.1), (5.2), (5.3), (5.4), \text{ and } (5.5),$$

$$0 \leq x_v \leq 1.$$

2: **for all** NF $n \in N$ **do**
3:     Sort $x_v, \forall v \in \bar{n}(n)$ decreasingly
4:     *counter* $= 0$
5:     **for all** VNF instance $v \in \bar{n}(n)$ **do**
6:         **if** *counter* $< A_n$ **then**
7:             *counter* $++, x_v = 1$
8:         **else**
9:             $x_v = 0$
10:        **end if**
11:    **end for**
12: **end for**
13: Take x_v into the **MILP**, and solve the resulted in LP to obtain the flow distribution λ_{vu}

The details of our relaxation-based algorithm are shown in Algorithm 7. To lower the computational complexity, the key idea of our algorithm is to relax the x_v into a real one within the range of $[0, 1]$. In this case, **Com-Min** can be rewritten into a linear programming (LP) problem as **Com-Min-LP** which can be solved in polynomial time by most solvers such as Matlab and Gurobi. Then, we try to solve this LP problem and obtain the x_v in float type as shown in line 1.

Next, the $x_v, \forall v \in \bar{n}(n)$ for NF n are sorted decreasingly and the first A_n x_vs are set to 1 while the others are set as 0 (lines 2–11). By such means, for each VNF, we have A_n instances placed in the network infrastructure. As long as the placement strategy is determined, we shall then find out how to balance the network flows between these network instances while reserving the network service semantics. Now, we obtain a VNF placement solution, i.e., the values of x_v, which are then taken into **Com-Min** and solve the resulted in **Com-Min-LP** to obtain the flow balancing solutions $\lambda_{uv}, \forall u, v \in V_v$.

5.2.4   Performance Evaluation

In this section, we compare our relax-based algorithm ("**RLX**") with the optimal result ("**OPT**") of **Com-Min** and the traditional one VNF algorithm ("**TOV**") [1], i.e., one VNF instance for each NF.

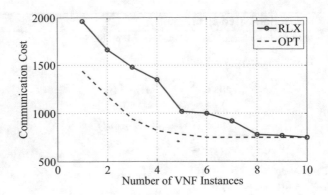

Fig. 5.3 The effect of the number of VNF instances

We use a network topology of $|D| = 10$ network nodes. The unit communication cost between each pair of network nodes H_{dp} is randomly set within the range of [1,10]. The NF DAG is also randomly generated with totally $|F| = 20$ network flows and $|N| = 10$ NFs. The source and consumer of each network flow are randomly located in this network and the flow rates Λ_{nm} are set in the range of [1, 10]. For each NF, we can have $A_n = 5$ VNF instances.

Commercial solver Gurobi[1] is used to solve our **Com-Min** and **Com-Min-LP** problems. We vary the settings of various parameters to see how they affect the communication cost in each experiment group.

First, we investigate how **OPT** and our **RLX** perform under different number of VNF instances from 2 to 10. It can be observed from Fig. 5.3 that the communication cost is shown as a decreasing function of the number of VNF instances A for both **OPT** and our **RLX**. The reason is that, in our **Com-Min** and **Com-Min-LP** formulation, constraint (5.1) limits the number of VNF instances for each NF and the corresponding network flow balancing. More VNF instances provide more optimization space for network flow balancing and can significantly lower the communication cost. One interesting phenomena is that the results of **OPT** and our **RLX** converge after $A_n = 8$ and become the same when $A = 10$. This is because with the increase of A_n, most NFs have enough numbers of VNF instances and further increasing A_n will not significantly reduce the communication cost. When A_n is large enough, e.g., $A_n = |D| = 10$, VNF instances can be placed in all network nodes and the network flows can be distributed among all them without any constraint. That is to say, constraint (5.1) cannot affect the network flow balancing any more. In this case, **Com-Min** and **Com-Min-LP** are equivalent and give the same communication cost.

Then, we vary the upper bound of number of hops from 1 to 10 and compare the results of the **OPT**, **RLX**, and **TOV** in Fig. 5.4. The communication costs of

---

[1]http://www.gurobi.com.

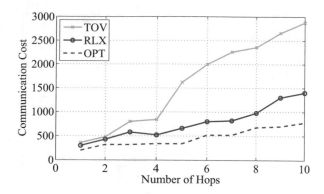

Fig. 5.4 The effect of the number of hops

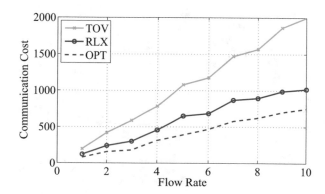

Fig. 5.5 The effect of the network flow rate

all three algorithms increase with upper bound of the number of hops. It is simply because with the same network flows and network topology, larger number of hops will lead to larger total communication cost according to (5.6). Therefore, the results of all three algorithms raise. Similarly, Figs. 5.5 and 5.6 show the results of all three algorithms increase with the number of NFs and number of network flows varying from 1 to 10 and 10 to 20, respectively.

Finally, Fig. 5.7 presents the results of all three algorithms as the number of NFs varying from 4 to 20. It can be obtained that the cost first increases and then converges. On one hand, all algorithms will try to place new VNF instances together with existing ones to lowering the communication cost between network nodes. On the other hand, the number of network nodes is set to 10. As the number of NFs increases, more VNF instances will be placed in the same DC without extra communication cost. Therefore, all results converge. Nevertheless, the advantage of our **RLX** can always be observed under any settings by the fact that it outperforms **TOV** and approaches **OPT**.

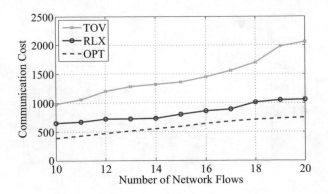

Fig. 5.6 The effect of the number of network flows

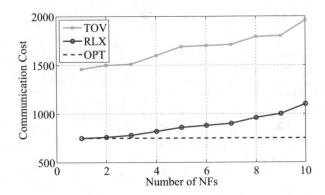

Fig. 5.7 The effect of the number of NFs

5.3 Online Traffic Scheduling in NFV

With the application of NFV technology, the roles of infrastructure provider and service provider can be clearly distinguished. Now, service providers mainly concern the operational cost for service provision as they rent both computation and communication resources from infrastructure providers to provide various network services in the form of network service chain [48, 49]. The recent development in cloud computing has resulted in a massive deployment of geo-distributed data centers interconnected by the Internet. In particular, the increasing communication requirements from various applications (e.g., Internet-of-Things streaming, etc.) have driven the deployment of edge servers collocated with users. This raises a new cloud computing concept called edge cloud [90], which is regarded as the next frontier for data centers. Edge cloud, with the advantage of user proximity, is an ideal platform to host network functions so as to provide network services faster, better, and cheaper.

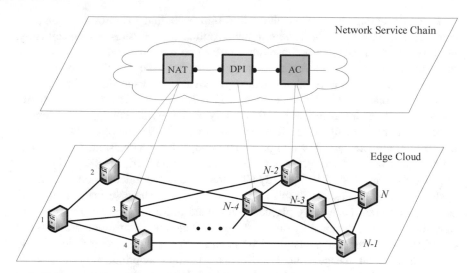

Fig. 5.8 Network service chain and edge cloud

To reserve the network service semantics, the packets of a network flow must go through a predetermined chain of network functions. For example, a typical firewall service may require a chain of network functions including a network address translation (NAT), a DPI, and an AC, as shown in Fig. 5.8. A network flow requiring firewall service shall sequentially go through these functions. Precisely speaking, the network flow must sequentially and completely visit all the edge servers hosting the corresponding network functions. Note that the communication cost between different pair of servers may vary. As a result, the network function placement will directly affect the packet scheduling and hence the overall communication cost. Therefore, how to explore the cost diversity feature of edge cloud for lowering the overall cost therefore naturally becomes one of the major concerns to network service providers.

Actually, network function management and packet scheduling (e.g., flow balancing) in NFV have already attracted much attention in the literature. For example, Moens et al. [73] study a network function placement problem on how to minimize the number of activated servers. Later, Cohen et al. [16] also investigate a function location problem with the goal of minimizing the overall cost including setup cost and distance cost. Addis et al. [1] further take the routing of network service chains across a carrier network into consideration. Although much effort has been already devoted to this area, we notice that existing studies usually assume a preknown network traffic demand, i.e., with prior knowledge. However, in practice, this is a luxurious assumption as the traffic demands usually vary dynamically, e.g., abruptly increasing or decreasing. As a result, these solutions may fail to orchestrate the network flows and manage the network functions at runtime according to the real-time network status.

In this section, we take an alternative approach that explores the queueing information available in the system to make online control decisions. Specifically, we leverage the Lyapunov optimization to design a dynamic control algorithm on packet scheduling, network function management, and resource allocation towards cost-efficient NFV in edge cloud. Although dynamic task scheduling has already been widely discussed in the literature, e.g., [21, 76, 79, 126], these existing studies are based on fixed infrastructure and focus only on the task scheduling. The problem becomes more challenging in NFV, because the task (i.e., packet) scheduling is correlated to the network function placement and activation decision. A packet can only be distributed to the server hosting the required network function. Moreover, the server resources shall be carefully allocated among the functions on it to ensure system stability. Therefore, in this section, we consider the case that the network functions can be dynamically placed or activated according to real-time network conditions. We systematically examine the cost efficiency optimization problem and establish a comprehensive analytical framework on the online control decision to NFV management in edge cloud. Following the Lyapunov optimization framework, the one-slot drift upper bound and the tradeoff between the queue backlog and the cost are derived. We propose a backpressure based online scheduling (BPOS) algorithm operating at runtime without any prior knowledge of future traffic demand.

5.3.1   System Model

In this paper, we consider an edge cloud infrastructure and a network service chain as shown in Fig. 5.8. The edge cloud is regarded as a distributed system with N interconnected edge servers. Let S denote the server set. The service chain consists of a set of network functions that must be sequentially visited. Assuming that there are M network functions, we denote the functions in the service chain as an index set $I = \{1, 2, \ldots, M\}$. A network flow requesting the service must go through all the required network functions defined in the service chain. In essence, the network service is deployed as an overlay in the edge cloud and all flow packets must sequentially go through the servers hosting the required network functions.

A server has a set of resources (e.g., CPU, disk, memory, etc.) that can be allocated to the functions residing on it. We specially focus on the CPU resource (i.e., computation resource) as it is the main bottleneck to the performance of the network function running on a server. The computation resource capacity on a server $u \in S$ is denoted as C_u. Different functions have different computation resource requirements. We assume that, to process a packet, α_i computation resource is required for function i. Accordingly, whenever network function i is placed and activated on server u in a time slot, P_u^i is billed.

Any two servers u and v can communicate with each other with a communication capacity denoted as C_{uv}. The communication cost between a pair of servers is charged in a "pay-as-you-go" manner according to the link utilization, i.e., traffic flow volume. The unit communication cost between servers u and v is represented by P_{uv}. Specially, we set the unit communication cost and capacity between the same server as 0 and infinity, respectively, i.e., $P_{uu} = 0$, $C_{uu} = \infty$, $\forall u \in S$.

A network flow requesting the network service shall first arrive at a front-end proxy server. A set of proxies are deployed in the edge cloud to collect the network service requests and behave as admission controllers and load balancers. The location (i.e., server in the cloud) of a proxy is predetermined. Without misunderstanding, we denote the proxy as "function 0," i.e., $i = 0$. We consider a time-slotted system where new service requests arrive at a proxy in server u according to a random arrival process $A_u(t)$. The arrival process is individually and independently distributed and therefore has no relationship between existing workload in the system. We do not have any knowledge on the statistical characteristics of the time-varying rate $A_u(t)$ as well.

5.3.2 Control Decision and Problem Statement

From the perspective of network service provider, it is significant to minimize the cost for resource renting from infrastructure providers. According to [64], a network function image is usually in a size of few megabytes and can be activated in few milliseconds. As a result, it is possible to provide network service in an on-demand way at runtime. Upon receiving a packet, the proxy first decides which server with the first function (i.e., function 1) it shall be distributed to. Thereafter, the packet proceeds along the functions defined in the service chain until the last function (i.e., function M) and similar control decision shall be made on each visited server. We are interested in finding an online packet scheduling and network function management policy that minimizes the overall cost and stabilizes the system for a service request rate under the service capacity of the system.

5.3.3 Cost-Efficient Problem Formulation

The service capacity for a service chain on an edge cloud is obtained by maximizing the utilization of all available resources. For a feasible request rate, it is desirable to carefully manage the available resources for overall cost minimization. Without any prior knowledge, all the control decisions shall be made at runtime. In other words, the control decision must be able to adapt to the time-varying request rates as well as the backlog conditions.

Let $\lambda_{uv}^i(t)$ denote the packets processed by server u hosting function i and distributed to server v hosting function $i + 1$ (omitted in the notation for brevity) at time t. The total number of packets that can be distributed is limited by the service rate, i.e.,

$$\sum_{v \in S} \lambda_{uv}^{i}(t) \le \mu_{u}^{i}(t), \forall u \in S, \tag{5.7}$$

where $\mu_{u}^{i}(t)$ is the service rate of function i on server u at time t. As server u may host a number of different functions at the same time, the total computation resources required by all functions are limited by the computation capacity, i.e.,

$$\sum_{i \in I} \alpha^{i} \mu_{u}^{i}(t) \le C_{u}, \forall u \in S. \tag{5.8}$$

Besides, the communication link between servers u and v is also shared by the flows between the hosting functions. The communication capacity limits the total flow volume as

$$\sum_{i \in I} \lambda_{uv}^{i}(t) \le C_{uv}, \forall u, v \in S. \tag{5.9}$$

Let $Q_{v}^{i}(t)$ denote the queue backlog length accumulated at server v for function i. The queueing dynamics for function i on server v can be described by

$$Q_{v}^{i}(t+1) = \max[Q_{v}^{i}(t) - \sum_{w \in S} \lambda_{vw}^{i}(t), 0] + A_{v}^{i}(t), \tag{5.10}$$

where

$$A_{v}^{i}(t) = \begin{cases} A_{v}(t), \text{ if } i = 0, \\ \sum_{u \in S} \lambda_{uv}^{i-1}(t), \text{ otherwise.} \end{cases} \tag{5.11}$$

The request arrival rate at a queue except the proxy server is determined by the packet scheduling decision at all its precedent queues. To say the system is stable, the following criteria must be met:

$$\bar{Q}_{v} \triangleq \limsup_{t \to \infty} \frac{1}{t} \mathbb{E}\{Q_{v}^{i}(t)\} < \infty, \forall i \in I, v \in S. \tag{5.12}$$

A server u can process the packets requesting function i if and only if function i is placed and activated on it. Let $x_{u}^{i}(t), \forall i \in I, u \in S$ denote such status, i.e.,

$$x_{u}^{i}(t) = \begin{cases} 1, \text{ function } i \text{ is activated on server } u \text{ at time } t, \\ 0, \text{ otherwise.} \end{cases} \tag{5.13}$$

We have

$$\frac{\mu_{u}^{i}(t)}{A} \le x_{u}^{i}(t) \le A\mu_{u}^{i}(t), \forall i \in I, u \in S, \tag{5.14}$$

where A is an arbitrary large number. Constraint (5.14) implies that whenever $\mu_u^i(t) > 0$, $x_u^i(t) \equiv 1$; otherwise, $x_u^i(t) \equiv 0$.

The communication cost is charged in a "pay-as-you-go" manner according to the actual link usage. The total communication cost $l(t)$ at time t can be calculated as:

$$l(t) = \sum_{i \in I} \sum_{u \in S} \sum_{v \in S} \lambda_{uv}^i(t) P_{uv}. \tag{5.15}$$

As activating a function i on server u requires a cost P_u^i, the total computation cost therefore can be expressed as:

$$d(t) = \sum_{i \in I} \sum_{u \in S} x_u^i(t) P_u^i. \tag{5.16}$$

Summing up both the computation cost and communication cost, we obtain the overall cost as:

$$P(t) = l(t) + d(t). \tag{5.17}$$

Our objective is to minimize the overall operational cost, including both communication and computation cost. Taking all the issues discussed as above, the cost efficiency optimization problem can be formulated as follows:

$$\min : \lim_{T \to \infty} \frac{1}{T} \sum_{t=0}^{T-1} P(t)$$
$$\text{s.t.} : (5.7)–(5.14). \tag{5.18}$$

Due to the burst and fluctuation of the packet rates, it is difficult to predict the workload at runtime. Therefore, how to manage the network functions in an edge cloud with the consideration of both resource allocation and packet scheduling to achieve the goal of cost efficiency is a challenging problem.

5.3.4 Cost-Efficient Online Scheduling Algorithm

In this section, we use the framework of Lyapunov optimization to develop an online scheduling algorithm for the cost efficiency optimization problem.

5.3.4.1 Lyapunov Drift Bound

The packet scheduling and resource allocation decision directly act on the queue backlog, i.e., congestion state, of a network function on a server. For example, distributing too many packets to a function allocated with little resource may incur long queue backlog or even system instability. To measure the aggregated queue backlog, we define a Lyapunov function as follows:

$$L(Q(t)) \triangleq \frac{1}{2} \sum_{u \in S} \sum_{i \in I} Q_u^i(t)^2 \tag{5.19}$$

and the corresponding one-slot conditional Lyapunov drift function describing the expected change in one time slot as:

$$\Delta(Q(t)) \triangleq \mathbb{E}\{[L(Q(t+1)) - L(Q(t))|Q(t)\}, \tag{5.20}$$

by which we are able to measure the congestion or the queue's stability.

Next, we calculate the upper bound of $\Delta(Q(t))$ subject to the resource capacity constraints.

Lemma 5.1 *Under any control decision without violating the resource capacity constraints, the Lyapunov drift $\Delta(Q(t))$ is bounded as:*

$$\Delta Q(t) \leq MNW^2 + \mathbb{E}\left\{\sum_{u \in S} \sum_{i \in I}\left[Q_u^i(t)(A_u^i(t) - \sum_{v \in S}\lambda_{uv}^i(t))|Q(t)\right]\right\}, \tag{5.21}$$

where $W = \max\{C_u/\alpha_i, C_{uv}\}, \forall u, v \in S, i \in I$.

Proof According to (5.19) and (5.20), we have

$$\Delta Q(t) = \mathbb{E}\{[L(Q(t+1)) - L(Q(t))|Q(t)]\}$$

$$= \mathbb{E}\left\{\frac{1}{2}\sum_{u \in S}\sum_{i \in I}Q_u^i(t+1)^2 - \frac{1}{2}\sum_{u \in S}\sum_{i \in I}Q_u^i(t)^2|Q(t)\right\}$$

$$= \frac{1}{2}\mathbb{E}\left\{\sum_{u \in S}\sum_{i \in I}\left[Q_u^i(t+1)^2 - Q_u^i(t)^2\right]|Q(t)\right\}$$

$$\leq \frac{1}{2}\mathbb{E}\left\{\sum_{u \in S}\sum_{i \in I}\left[A_u^i(t)^2 + \sum_{v \in S}\lambda_{uv}^i(t)^2\right.\right. \tag{5.22}$$

$$\left.\left. +2Q_u^i(t)(A_u^i(t) - \sum_{v \in S}\lambda_{uv}^i(t))|Q(t)\right]\right\}$$

$$= \frac{1}{2}\mathbb{E}\left\{\sum_{u \in S}\sum_{i \in I}\left[A_u^i(t)^2 + \sum_{v \in S}\lambda_{uv}^i(t)^2\right]\right\}$$

$$+ \mathbb{E}\left\{\sum_{u \in S}\sum_{i \in I}\left[Q_u^i(t)(A_u^i(t) - \sum_{v \in S}\lambda_{uv}^i(t))|Q(t)\right]\right\}.$$

Combining constraints (5.7) and (5.8), leads to

$$\sum_{v \in S} \sum_{i \in I} \alpha^i \lambda_{uv}^i(t) \le C_u, \forall u \in S. \tag{5.23}$$

From (5.23) and (5.9), we have

$$\sum_{v \in S} \lambda_{uv}^i(t) \le \max\{C_u/\alpha_i, C_{uv}\}, \forall u, v \in S, i \in I. \tag{5.24}$$

By defining

$$W \triangleq \max\{C_u/\alpha_i, C_{uv}\}, \forall u, v \in S, i \in I, \tag{5.25}$$

and taking it into (5.22), we further have

$$
\begin{aligned}
\Delta Q(t) &\le \frac{1}{2} \mathbb{E} \left\{ \sum_{u \in S} \sum_{i \in I} \left[A_u^i(t)^2 + \sum_{v \in S} \lambda_{uv}^i(t)^2 \right] \right\} \\
&\quad + \mathbb{E} \left\{ \sum_{u \in S} \sum_{i \in I} \left[Q_u^i(t)(A_u^i(t) - \sum_{v \in S} \lambda_{uv}^i(t)) \right] \right\} \\
&= \frac{1}{2} MNW^2 + \frac{1}{2} MNW^2 \\
&\quad + \mathbb{E} \left\{ \sum_{u \in S} \sum_{i \in I} \left[Q_u^i(t)(A_u^i(t) - \sum_{v \in S} \lambda_{uv}^i(t)) | Q(t) \right] \right\} \\
&= MNW^2 + \mathbb{E} \left\{ \sum_{u \in S} \sum_{i \in I} \left[Q_u^i(t)(A_u^i(t) - \sum_{v \in S} \lambda_{uv}^i(t)) | Q(t) \right] \right\}.
\end{aligned}
\tag{5.26}
$$

∎

5.3.4.2   Optimality Analysis

Lemma 5.1 provides us a tool to keep the queue stable subject to the resource capacity constraints. In this section, we further take the overall operational cost into consideration and analyze the optimality in terms of the tradeoff between queue backlog and cost.

In practice, for a network function residing on a server, the packet arrival rate may exceed the service rate determined by the allocated resource in a time slot. Hence, for each queue, these must exist a real number $\epsilon > 0$ with the following property:

$$\mathbb{E}\left\{ A_u^i(t) - \sum_{v \in S} \lambda_{uv}^i | Q(t) \right\} \leq -\epsilon, \forall u \in S. \tag{5.27}$$

As a result, the drift bound can be rewritten as:

$$\Delta Q(t) \leq MNW^2 - \epsilon \sum_{u \in S} \sum_{i \in I} Q_u^i(t). \tag{5.28}$$

In order to the take the overall cost into consideration, by taking advantage of the Lyapunov optimization framework, we further define a drift-plus-penalty expression as:

$$\Delta Q(t) + V\mathbb{E}\{P(t)|Q(t)\}, \tag{5.29}$$

where V is a control parameter for balancing the tradeoff between queue backlog and cost. Adding $V\mathbb{E}\{P(t)|Q(t)\}$ to both sides of (5.28) yields a bound on the drift-plus-penalty:

$$
\begin{aligned}
&\Delta(Q(t)) + V\mathbb{E}\{P(t)|Q(t)\} \\
&\leq MNW^2 - \epsilon \sum_{u \in S} \sum_{i \in I} Q_u^i(t) + V\mathbb{E}\{P(t)|Q(t)\} \\
&\leq MNW^2 - \epsilon \sum_{u \in S} \sum_{i \in I} Q_u^i(t) + VP^*,
\end{aligned} \tag{5.30}
$$

where P^* denotes the optimal overall cost over T time slots and is defined as:

$$P^* = \frac{1}{T} \sum_{t=0}^{T-1} P^*(t). \tag{5.31}$$

Summing both the left-hand side and right-hand side in (5.30) over $t \in \{0, 1, \ldots, T-1\}$, we obtain

$$
\begin{aligned}
&\mathbb{E}\{L(Q(T))\} - \mathbb{E}\{L(Q(0))\} + V \sum_{t=0}^{T-1} \mathbb{E}\{P(t)\} \\
&\leq MNW^2 T + VTP^* - \epsilon \sum_{t=0}^{T-1} \sum_{u \in S} \sum_{i \in I} \mathbb{E}\left\{ Q_u^i(t) \right\}.
\end{aligned} \tag{5.32}
$$

By rearranging and relaxing (5.32), the following two inequalities are derived:

$$\frac{1}{T} \sum_{t=0}^{T-1} \mathbb{E}\{P(t)\} \leq \frac{MNW^2}{V} + P^* + \frac{\mathbb{E}\{L(Q(0))\}}{VT} - \frac{\mathbb{E}\{L(Q(T))\}}{VT} \tag{5.33}$$

note that, $L(Q(T)) \geq 0$, so we have

$$\frac{1}{T}\sum_{t=0}^{T-1}\mathbb{E}\{P(t)\} \leq \frac{MNW^2}{V} + P^* + \frac{\mathbb{E}\{L(Q(0))\}}{VT} \tag{5.34}$$

and

$$\frac{1}{T}\sum_{t=0}^{T-1}\sum_{u\in S}\sum_{i\in I}\mathbb{E}\{Q_u^i(t)\} \leq \frac{MNW^2 + VP^* - \frac{1}{T}V\sum_{t=0}^{T-1}\mathbb{E}\{P(t)\}}{\epsilon} + \frac{\mathbb{E}\{L(Q(0))\}}{\epsilon T}. \tag{5.35}$$

Making $T \to \infty$, (5.34) and (5.35) yield

$$\lim_{T\to\infty}\frac{1}{T}\sum_{t=0}^{T-1}\mathbb{E}\{P(t)\} \leq \frac{MNW^2}{V} + P^*, \tag{5.36}$$

and

$$\lim_{T\to\infty}\frac{1}{T}\sum_{t=0}^{T-1}\sum_{n=1}^{N}\sum_{i=1}^{I}Q_n^i(t) \leq \frac{MNW^2 + VP^*}{\epsilon}. \tag{5.37}$$

The bounds (5.36) and (5.37) show an $[O(1/V), O(V)]$ backlog-cost tradeoff. An arbitrarily large V makes the backlog-cost ratio arbitrarily small, implying that the time average overall cost is arbitrarily close to the optimum P^* and the average queue backlog bound is $O(V)$. On the contrary, setting a small V leads to a biased queue backlog and a large overall cost. By Little's theorem, the average queue backlog is proportional to the average process delay on that queue. The tradeoff is actually pushed towards optimality with a tradeoff between the end-to-end delay and overall cost.

5.3.4.3   Online Algorithm Design

According to the backpressure algorithm, the problem formulated in (5.18) can be solved in a greedy and distributed manner on each server at each time slot $t \in \{0, 1, \cdots, T-1\}$ according to the observed queue statuses $Q(t)$ and available resource capacity $C_u, C_{uv}, \forall u, v \in S$. Basically, backpressure algorithm intends to minimize the backlog in each time slot for each server u. At runtime, the backlog for function $i \in I$ on a node $u \in S$ is reflected by the packets processed and distributed to its descendant nodes. Minimizing the backlog is equivalent to maximizing the output rate, i.e.,

---

Algorithm 8 Backpressure based online scheduling algorithm

1: **for** each time slot $t \in \{0, 2, \cdots, T - 1\}$ **do**
2:     calculate the queue backlog and available resource capacities
3:     solve (5.40) to get the values of $\lambda_{uv}^i, \mu_u^i, x_u^i, \forall u, v \in S, i \in I$
4:     allocate resource and process the queueing packets accordingly
5:     distribute the processed packets to the descendant servers hosting with the next function
6: **end for**

$$\max : \sum_{i \in I} \sum_{v \in S} \lambda_{uv}^i(t). \tag{5.38}$$

However, arbitrarily maximizing the output rate may incur a high cost. In order to balance the backlog and cost, we further take the cost into consideration. According to (5.15) and (5.16), the total cost of server u at time t for all functions residing on it can be calculated as:

$$P_u(t) = \sum_{i \in I} \sum_{v \in S} \lambda_{uv}^i(t) P_{uv} + \sum_{i \in I} x_u^i(t) P_u^i. \tag{5.39}$$

Following (5.29), the objective for server u at time slot t shall be transformed to $\max : \sum_{v \in S} \lambda_{uv}^i(t) - V P_u(t)$, where V is the control parameter introduced in (5.29). Taking all the resource capacity constraints into consideration, we obtain an ILP formulation that shall be solved on each server $u \in S$ at each time slot t as follows.

$$\max : \sum_{v \in S} \lambda_{uv}^i(t) - V P_u(t)$$

$$\text{s.t.} : \sum_{v \in S} \alpha^i \lambda_{uv}^i(t) \leq \mu_u^i(t)$$

$$\sum_{i \in I} \mu_u^i(t) \leq C_u \tag{5.40}$$

$$\sum_{i \in I} \lambda_{uv}^i(t) \leq C_{uv}, \forall v \in S$$

$$\frac{\mu_u^i(t)}{A} \leq x_u^i(t) \leq A \mu_u^i(t), \forall i \in I$$

Accordingly, we design our cost-efficient backpressure based online scheduling (BPOS) algorithm, which is executed online at each server in runtime. The pseudo code is given in Algorithm 8.

For each server u at the beginning of each time slot t, we first calculate the system states related to u, including queue backlog and the capacities of various resources in line 2. Taking all these observations into (5.40), we next try to obtain the decisions on the resource allocation μ_u^i and packet scheduling λ_{uv}^i in line 3. Note that, the

ILP shown in (5.40) only involves M binaries $x_u^i(t)$ for all functions and can be solved in a time-efficient manner using the branch and bound heuristic method. The details for solving the ILP are omitted for brevity. Then, we allocate the resource, process the queueing packets, and distribute the processed ones to the descendant accordingly in lines 4 and 5. Such process proceeds at each server until the end of system operation at time slot $T - 1$.

5.3.5 Performance Evaluation

5.3.5.1 Experiment Setting

To explore the performance of our BPOS algorithm, a simulator exactly following the system model as described in Sect. 4.3.1 has been implemented. We simulate under the setting that a number of user requests are received by the proxy servers at each time slot. All requests must first go through a network service deployed in the edge cloud. The edge cloud has 15 servers and the network service chain has five network functions. A real-world network request trace from the BRAIN research network in Berlin is adopted to simulate the user requests. The user request rate along timeline is plotted in Fig. 5.9a. The default resource capacities of the

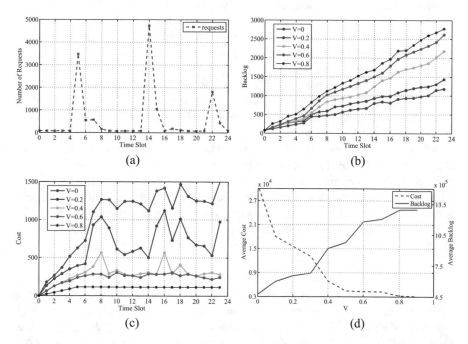

Fig. 5.9 Trace-driven evaluation results on the performance of BPOS. (**a**) The number of requests for trace "BRAIN" at different time slots. (**b**) Queue backlog at different time slots. (**c**) Overall cost at different time slots. (**d**) Time-averaged backlog and cost under different values of V

servers in the edge cloud are set as follows. Each server is with a total computation capacity 20. The communication capacity between a pair of servers is 50. The unit communication cost is randomly assigned in range $[1, 5]$ to simulate the cost diversity.

We apply two different strategies as baseline algorithms to show the efficiency of our BPOS algorithm. The first method called "All-in-One" tries to place all virtual functions to one server such that no inter-server communication is needed and hence communication cost can be minimized. The second one "One-in-One" allows only one function on a server such that all computation resources can be allocated to the hosted function for fast queue evacuation.

5.3.5.2   Evaluation and Analysis

As we have known, the value of parameter V controls a tradeoff between the queue backlog and the cost. We first show how the value of V affects the queue backlog and the cost to verify the correctness and efficiency of our design.

To check how the queue backlog changes at runtime during each time slot, we first plot the overall queue backlog along the timeline on different values of control parameter V in Fig. 5.9b. We first notice that the queue backlog increases along the timeline. As new requests continuously arrive at the proxies, the to-be-processed packets consequently accumulate at each server resulting in the increased queue backlog. Another phenomenon deserved attention is that the increasing trends vary on the values of V. The larger value of V, the faster increasing of the queue backlog. This validates our analysis that smaller V biases more on the queue backlog and implies faster packet processing.

Next, we plot the instant total cost along the timeline in Fig. 5.9c. We see that the cost fluctuates over the timeline under different values of V except $V = 0.8$. When $V = 0.8$, the cost is comparatively stable over the timeline. Obviously, the cost fluctuation is due to the time-varying request rates as shown in Fig. 5.9a. As we have known, with smaller V, our BPOS algorithm seeks to explore more resources available in the system for faster queue evacuation. Therefore, the cost changes dramatically following the request variation. However, when V becomes larger, BPOS tries to use less resources provided that the queue stability requirement is not violated and hence the cost becomes comparatively stable with a larger V. Moreover, larger V also indicates lower cost at each time slot. As we can see from Fig. 5.9c, $V = 0.8$ always achieves the lowest cost among all other values of V. This also verifies the philosophy of our BPOS design.

To summarize the effect of V to the backlog-cost tradeoff, we further present the average backlog and cost under different values of V in Fig. 5.9d. We can clearly observe that the average backlog and cost show as increasing and decreasing functions of V, respectively. This phenomenon is consistent to our design principle. With a smaller value of V, we bias on the system performance and hence a shorter backlog shall be observed. On the contrary, we may emphasize the cost with a larger

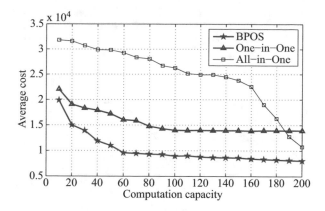

Fig. 5.10 Overall cost under different computation capacities

value of V. By tuning the value of V, we are able to achieve the desired tradeoff between the backlog and the overall cost.

Next, we compare the average cost achieved by our algorithm against "All-in-One" and "One-in-One" strategies. In this group of experiments, we set the control parameter $V = 0.3$ according to the results from Fig. 5.9d.

We first vary the computation capacity on each server from 10 to 200 and check the time-averaged cost. The evaluation results are presented in Fig. 5.10. We notice that the overall cost decreases with the increasing of computation capacity. This is attributed to the fact that less functions need to be activated to ensure the queue stability and less computation cost shall be required with larger computation capacity on each server. We also observe that our BPOS algorithm exhibits much advantage over the competitors under any computation capacity. This verifies the high efficiency of BPOS algorithm as it can well balance the computation and communication resources to achieve the cost efficiency goal.

The effect of the communication capacity to the overall cost is plotted in Fig. 5.11 where the communication capacity varies from 10 to 200. We notice that the cost shows as a decreasing function of the communication capacity. With larger communication capacity, the links with less unit communication cost are utilized. Besides, thanks to faster queue evacuation ability, less functions shall be activated to ensure the queue stability. In addition, larger communication capacity indicates that the system is less constrained by the communication resource but by the computation resource. As a result, we shall see that the cost converges after the communication capacity reaches 50 by BPOS algorithm. Nevertheless, once again, we can see that our algorithm much outperforms both "All-in-One" and "One-in-One" under any communication capacity.

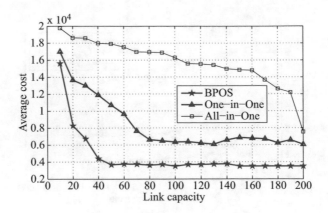

Fig. 5.11 Overall cost under different communication capacities

5.4   Conclusion

By applying the newly emerging NFV technology, the flexibility on network management is much approved as the VNF instances can be freely placed in the network. In this chapter, we first investigate a communication cost minimization problem with joint consideration of VNF instance placement and network flow balancing, without violating the predetermined network service semantics. We formulate the VNF placement problem into an MILP form and propose a low-complexity heuristic algorithm by relaxing the integer variables in the MILP formulation. Then, we address the problem of online packet scheduling, function management, and resource allocation towards cost-efficient NFV in edge cloud. For any feasible network service request rates under the service capacity, we investigate how to minimize the cost of network service provision. The Lyapunov optimization framework is explored to analyze the essence of the online scheduling. The drift bound as well as the tradeoff between the queue backlog and the cost are derived in closed forms. A backpressure based online scheduling algorithm is proposed and evaluated. Our work provides a valuable insight into online packet scheduling, resource allocation, and function management for NFV.

Chapter 6
Conclusion and Future Research Directions

6.1 Concluding Remarks

In this book, we introduce the concept and architecture of software defined systems (SDS). The core enabling technologies, including software defined front-end devices (sensors, IoT devices), software defined access networks (e.g., cognitive radio, CRAN), software defined core networks (e.g., SDN, NFV), software defined storage and computing (e.g., microservice). These technologies jointly enable the programmers or system administrators escape from the heavy reliance on hardware. With SDS, all sensing, communication, computation, and storage resources can be managed in a software-defined way, much promoting the system flexibility and lowering the barrier for information system innovation. SDS brings not only new opportunities, but also new challenges, in resource management and optimization. With the advent of new information technologies, corresponding new resource management and optimization algorithms shall be designed to cater for the indistinguishable characteristics of these new technologies.

Therefore, after introducing the basic concept of SDS, we then introduce some newly proposed resource management and optimization algorithms designed for SDS. In software defined sensor networks, we introduce a performance efficiency online task scheduling in multi-task software defined sensors to deal with various unpredictable network dynamics at runtime. In software defined access networks, we focus on cloud-radio access networks (CRAN) and introduce two algorithms to improve the network energy efficiency without violating the performance efficiency requirements, for remote radio head (RRH) and base-band unit (BBU) management, respectively. In software defined core networks, we first introduce two algorithms for forwarding rule management and controller placement, respectively, to deal with the problems introduced by the separation of control plane and data plane in software defined networks (SDNs). Then, we discuss the virtualized network function

(VNF) management in network function virtualization (NFV) empowered networks, under the assumption of known and unknown flow request rates, respectively.

6.2 Potential Future Works

SDS shall be the development trend of information systems. According to the related studies as well as the enabling technologies introduced in this book, it is obvious that such trend is already on the road. While, this book mainly focuses on the resource management algorithm design in SDS. With the promising vision of SDS, there are many potential future works for the further development, and practical deployment and application, at least, but not limited to.

- *Programming framework:* To pursue the wide adoption and application of SDS, one of the most important issues is the programming framework. Previous information technology development experiences have already told us the prosperousness of any new technology cannot be achieved without a promising programming framework, e.g., Hadoop to cloud computing. This shall be also the same to SDS, which requires a unified and friendly framework that can manage all the resources in SDS, ranging from the front-end devices like sensors and actuators, to the communication networks, and to the back-end servers and services.
- *Intelligent management:* The success of AlphaGo ignites the third wave of artificial intelligence (AI) technology. It has already widely recognized that AI is not only a charming information technology but also a promising technology to information system, especially on the resource management and optimization. The software definability of SDS naturally fits the application capability of many AI technologies. For example, we can apply reinforcement learning to plan the routing path of network flows by exploring SDN. Far beyond that, all the resources can be efficiently and intelligently controlled by AI empowered controller to pursue the goal of from software defined system to self-running systems.
- *Industrial standards:* To make SDS seamlessly integrate with the AI empowered controller, which may be development by different people or companies, it is first required that all the resources shall have unified and standard interfaces. Just like the case in SDN, without such interface, it is hard, or even impossible, to make all the communication devices interconnected and controlled by a logically centralized controller. On the other hand, commercial drive is always one of the main forces to the successful deployment and application of new technologies. Recognized industrial standards are highly needed to improve the openness and interoperability of SDS.

- *Cross-domain security:* By SDS, all the sensing, communication, computation, and storage resources will be shared in a transparent way to the system developers, to whom security is always one of the main concerns. Unlike traditional information systems, SDS breaks the boundary in different domains and this further exaggerates the security challenges. For example, we shall not only concern the security of data storage, but shall take care of the data security throughout its life cycle, from sensing to storage, as all the resources that support the data generation, transferring, processing, and storage are shared by different developers.

References

1. Bernardetta Addis, Dallal Belabed, Mathieu Bouet, and Stefano Secci. Virtual Network Functions Placement and Routing Optimization. In *Proceedings of the 4th International Conference on Cloud Networking (CloudNet)*, pages 171–177. IEEE, 2015.
2. Ian F. Akyildiz, Pu Wang, and Shih Chun Lin. SoftAir: A Software Defined Networking Architecture for 5G Wireless Systems. *Computer Networks*, 85:1–18, 2015.
3. Gunther Auer, Vito Giannini, Claude Desset, Istvan Godor, Per Skillermark, Magnus Olsson, Muhammad Ali Imran, Dario Sabella, Manuel J Gonzalez, Oliver Blume, et al. How Much Energy is Needed to Run a Wireless Network? *IEEE Wireless Communications*, 18(5):40–49, 2011.
4. Armin Balalaie, Abbas Heydarnoori, and Pooyan Jamshidi. Microservices architecture enables devops: Migration to a cloud-native architecture. *Ieee Software*, 33(3):42–52, 2016.
5. Fouad Benamrane, Redouane Benaini, et al. An east-west interface for distributed SDN control plane: Implementation and evaluation. *Computers & Electrical Engineering*, 57:162–175, 2017.
6. Kevin Benton, L Jean Camp, and Chris Small. Openflow vulnerability assessment. In *Proceedings of the second ACM SIGCOMM workshop on Hot topics in software defined networking*, pages 151–152. ACM, 2013.
7. Md Zakirul Alam Bhuiyan, Guojun Wang, Jiannong Cao, and Jie Wu. Sensor placement with multiple objectives for structural health monitoring. *ACM Transactions on Sensor Networks (TOSN)*, 10(4):68, 2014.
8. Daniel Bienstock. Computational study of a family of mixed-integer quadratic programming problems. *Mathematical Programming*, 74(2):121–140, 1996.
9. ONF Solution Brief. Openflow-enabled SDN and network functions virtualization. *Open Netw. Found*, 17:1–12, 2014.
10. Emmanuel J Candes, Michael B Wakin, and Stephen P Boyd. Enhancing Sparsity by Reweighted ℓ 1 Minimization. *Journal of Fourier analysis and applications*, 14(5):877–905, 2008.
11. Jiming Chen, Kejie Cao, Keyong Li, and Youxian Sun. Distributed sensor activation algorithm for target tracking with binary sensor networks. *Cluster Computing*, 14(1):55–64, 2011.
12. Mung Chiang and Tao Zhang. Fog and IoT: An overview of research opportunities. *IEEE Internet of Things Journal*, 3(6):854–864, 2016.
13. Margaret Chiosi, Don Clarke, Peter Willis, Andy Reid, James Feger, Michael Bugenhagen, Waqar Khan, Michael Fargano, Chunfeng Cui, Hui Deng, et al. Network functions virtualisation: An introduction, benefits, enablers, challenges and call for action. In *SDN and OpenFlow World Congress*, volume 48. sn, 2012.

14. NM Mosharaf Kabir Chowdhury and Raouf Boutaba. A survey of network virtualization. *Computer Networks*, 54(5):862–876, 2010.

15. David Clark. The design philosophy of the darpa internet protocols. *ACM SIGCOMM Computer Communication Review*, 18(4):106–114, 1988.

16. R. Cohen, L. Lewin-Eytan, J. S. Naor, and D. Raz. Near optimal placement of virtual network functions. In *2015 IEEE Conference on Computer Communications (INFOCOM)*, pages 1346–1354, April 2015.

17. Thomas H Cormen, Charles E Leiserson, Ronald L Rivest, Clifford Stein, et al. *Introduction to algorithms*, volume 2. MIT Press Cambridge, 2001.

18. Richard Cziva, Simon Jouet, and Dimitrios P Pezaros. Gnfc: Towards network function cloudification. In *2015 IEEE conference on network function virtualization and software defined network (NFV-SDN)*, pages 142–148. IEEE, 2015.

19. Binbin Dai and Wei Yu. Sparse Beamforming and User-centric Clustering for Downlink Cloud Radio Access Network. *IEEE Access*, 2:1326–1339, 2014.

20. Mou Dasgupta and GP Biswas. Design of multi-path data routing algorithm based on network reliability. *Computers & Electrical Engineering*, 38(6):1433–1443, 2012.

21. Apostolos Destounis, Georgios S. Paschos, and Iordanis Koutsopoulos. Streaming Big Data meets Backpressure in Distributed Network Computation. page 24, jan 2016.

22. Advait Dixit, Fang Hao, Sarit Mukherjee, TV Lakshman, and Ramana Kompella. Towards an elastic distributed SDN controller. *ACM SIGCOMM computer communication review*, 43(4):7–12, 2013.

23. Dmitry Drutskoy, Eric Keller, and Jennifer Rexford. Scalable network virtualization in software-defined networks. *IEEE Internet Computing*, 17(2):20–27, 2012.

24. Jose Oscar Fajardo, Ianire Taboada, and Fidel Liberal. Radio-aware service-level scheduling to minimize downlink traffic delay through mobile edge computing. In *International Conference on Mobile Networks and Management*, pages 121–134. Springer, 2015.

25. Nick Feamster, Jennifer Rexford, and Ellen Zegura. The road to sdn. *Queue*, 11(12):20, 2013.

26. Nick Feamster, Jennifer Rexford, and Ellen Zegura. The road to sdn: an intellectual history of programmable networks. *ACM SIGCOMM Computer Communication Review*, 44(2):87–98, 2014.

27. K. Foerster, S. Schmid, and S. Vissicchio. Survey of consistent software-defined network updates. *IEEE Communications Surveys Tutorials*, 21(2):1435–1461, Secondquarter 2019.

28. Alan Ford, Costin Raiciu, Mark Handley, Sebastien Barre, Janardhan Iyengar, et al. Architectural guidelines for multipath TCP development. *IETF, Informational RFC*, 6182:2070–1721, 2011.

29. Armando Fox, Rean Griffith, Anthony Joseph, Randy Katz, Andrew Konwinski, Gunho Lee, David Patterson, Ariel Rabkin, and Ion Stoica. Above the clouds: A berkeley view of cloud computing. *Dept. Electrical Eng. and Comput. Sciences, University of California, Berkeley, Rep. UCB/EECS*, 28(13):2009, 2009.

30. Jean-Marie Garcia, Olivier Brun, and David Gauchard. Transient analytical solution of m/d/1/n queues. *Journal of applied probability*, pages 853–864, 2002.

31. H. Ghalwash and C. Huang. A QoS framework for SDN-based networks. In *2018 IEEE 4th International Conference on Collaboration and Internet Computing (CIC)*, pages 98–105, Oct 2018.

32. Anoop Ghanwani, D Krishnaswamy, RR Krishnan, P Willis, N Sriram, A Chaudhary, and F Huici. An analysis of lightweight virtualization technologies for NFV. *IETFInternet-Draftdraft-natara-jan-nfvrg-containersfor-nfv-03, July2016*, 2016.

33. Frédéric Giroire, Joanna Moulierac, and T Khoa Phan. Optimizing rule placement in software-defined networks for energy-aware routing. 2014.

34. Lin Gu and John A Stankovic. Radio-triggered wake-up for wireless sensor networks. *Real-Time Systems*, 29(2–3):157–182, 2005.

35. Lin Gu, Deze Zeng, Song Guo, Ahmed Barnawi, and Yong Xiang. Cost Efficient Resource Management in Fog Computing Supported Medical Cyber-Physical System. *IEEE Transactions on Emerging Topics in Computing*, 5(1):108–119, 2015.

36. Gurobi Optimization. Gurobi Optimizer Reference Manual, 2013.
37. Vu Nguyen Ha and Long Bao Le. Joint Coordinated Beamforming and Admission Control for Fronthaul Constrained Cloud-RANs. In *Global Communications Conference (GLOBECOM), 2014 IEEE*, pages 4054–4059. IEEE, 2014.
38. Ibrahim Abaker Targio Hashem, Ibrar Yaqoob, Nor Badrul Anuar, Salimah Mokhtar, Abdullah Gani, and Samee Ullah Khan. The rise of "big data" on cloud computing: Review and open research issues. *Information systems*, 47:98–115, 2015.
39. Soheil Hassas Yeganeh and Yashar Ganjali. Kandoo: A Framework for Efficient and Scalable Offloading of Control Applications. In *Proceedings of the first workshop on Hot topics in Software Defined Networks*, pages 19–24. ACM, 2012.
40. Hassan Hawilo, Abdallah Shami, Maysam Mirahmadi, and Rasool Asal. NFV: State of the art, challenges and implementation in next generation mobile networks (vEPC). *arXiv preprint arXiv:1409.4149*, 2014.
41. Simon Haykin. Cognitive radio: brain-empowered wireless communications. *IEEE journal on selected areas in communications*, 23(2):201–220, 2005.
42. Brandon Heller, Rob Sherwood, and Nick McKeown. The Controller Placement Problem. In *Proceedings of the First Workshop on Hot Topics in Software Defined Networks*, pages 7–12. ACM, 2012.
43. Mingyi Hong, Ruoyu Sun, Hadi Baligh, and Zhi-Quan Luo. Joint Base Station Clustering and Beamformer Design for Partial Coordinated Transmission in Heterogeneous Networks. *IEEE Journal on Selected Areas in Communications*, 31(2):226–240, 2013.
44. Fei Hu, Qi Hao, and Ke Bao. A survey on software-defined network and openflow: From concept to implementation. *IEEE Communications Surveys & Tutorials*, 16(4):2181–2206, 2014.
45. Yannan Hu, Wang Wendong, Xiangyang Gong, Xirong Que, and Cheng Shiduan. Reliability-aware Controller Placement for Software-Defined Networks. In *Integrated Network Management (IM 2013), 2013 IFIP/IEEE International Symposium on*, pages 672–675. IEEE, 2013.
46. Yun Chao Hu, Milan Patel, Dario Sabella, Nurit Sprecher, and Valerie Young. Mobile edge computing-a key technology towards 5g. *ETSI white paper*, 11(11):1–16, 2015.
47. Chi-Fu Huang and Yu-Chee Tseng. The coverage problem in a wireless sensor network. *Mobile Networks and Applications*, 10(4):519–528, 2005.
48. Huawei Huang, Song Guo, Jinsong Wu, and Jie Li. Service Chaining for Hybrid Network Function. *IEEE Transactions on Cloud Computing*, 2017.
49. Huawei Huang, Peng Li, Song Guo, Weifa Liang, and Kun Wang. Near-Optimal Deployment of Service Chains by Exploiting Correlations between Network Functions. *IEEE Transactions on Cloud Computing*, 2017.
50. Meitian Huang, Weifa Liang, and Song Guo. Throughput Maximization of Delay-Sensitive Request Admission via Virtualized Network Function Placements and Migrations. In *IEEE International Conference on Communications (ICC)*. IEEE, 2018.
51. Neeraj Jaggi, Sreenivas Madakasira, Sandeep Reddy Mereddy, and Ravi Pendse. Adaptive algorithms for sensor activation in renewable energy based sensor systems. *Ad Hoc Networks*, 11(4):1405–1420, 2013.
52. M. Jarschel, S. Oechsner, D. Schlosser, R. Pries, S. Goll, and P. Tran-Gia. Modeling and Performance Evaluation of an OpenFlow Architecture. In *Teletraffic Congress (ITC), 2011 23rd International*, pages 1–7, Sept 2011.
53. Yury Jimenez, Cristina Cervello-Pastor, and Aurelio J Garcia. On the Controller Placement for Designing a Distributed SDN Control Layer. In *Networking Conference, 2014 IFIP*, pages 1–9. IEEE, 2014.
54. Wolfgang John, Konstantinos Pentikousis, George Agapiou, Eduardo Jacob, Mario Kind, Antonio Manzalini, Fulvio Risso, Dimitri Staessens, Rebecca Steinert, and Catalin Meirosu. Research directions in Network Service Chaining. In *SDN4FNS 2013 – 2013 Workshop on Software Defined Networks for Future Networks and Services*, 2013.

55. Paul Kolodzy and Interference Avoidance. Spectrum policy task force. *Federal Commun. Comm., Washington, DC, Rep. ET Docket*, 40(4):147–158, 2002.
56. Diego Kreutz, Fernando MV Ramos, Paulo Verissimo, Christian Esteve Rothenberg, Siamak Azodolmolky, and Steve Uhlig. Software-defined networking: A comprehensive survey. *Proceedings of the IEEE*, 103(1):14–76, 2015.
57. Greg LaBrie. Software Defined Systems: 5 Major Benefits To The Enterprise. https:// blog.wei.com/software-defined-systems-5-major-benefits-to-the-enterprise/, 2016. [Online; accessed 15-July-2019].
58. Adrian Lara, Anisha Kolasani, and Byrav Ramamurthy. Network innovation using openflow: A survey. *IEEE communications surveys & tutorials*, 16(1):493–512, 2013.
59. Philip Levis, Sam Madden, Joseph Polastre, Robert Szewczyk, Kamin Whitehouse, Alec Woo, David Gay, Jason Hill, Matt Welsh, Eric Brewer, et al. TinyOS: An Operating System for Sensor Networks. In *Ambient Intelligence*, pages 115–148. Springer, 2005.
60. Dapeng Li, Walid Saad, Ismail Guvenc, Abolfazl Mehbodniya, and Fumiyuki Adachi. Decentralized Energy Allocation for Wireless Networks with Renewable Energy Powered Base Stations. *IEEE Transactions on Communications*, 63(6):2126–2142, 2015.
61. Jian Li, Jingxian Wu, Mugen Peng, and Ping Zhang. *IEEE Transactions on Wireless Communications*, 15(6):3880–3894, 2016.
62. Yunhao Liu, Yuan He, Mo Li, Jiliang Wang, Kebin Liu, and Xiangyang Li. Does wireless sensor network scale? a measurement study on greenorbs. *Parallel and Distributed Systems, IEEE Transactions on*, 24(10):1983–1993, 2013.
63. Tie Luo, Hwee-Pink Tan, and T.Q.S. Quek. Sensor OpenFlow: Enabling Software-Defined Wireless Sensor Networks. *IEEE Communications Letters*, 16(11):1896–1899, 2012.
64. Joao Martins, Mohamed Ahmed, Costin Raiciu, Vladimir Olteanu, Michio Honda, Roberto Bifulco, and Felipe Huici. Clickos and the art of network function virtualization. In *Proceedings of the 11th USENIX Conference on Networked Systems Design and Implementation*, pages 459–473. USENIX Association, 2014.
65. Nick McKeown, Tom Anderson, Hari Balakrishnan, Guru Parulkar, Larry Peterson, Jennifer Rexford, Scott Shenker, and Jonathan Turner. Openflow: enabling innovation in campus networks. *ACM SIGCOMM Computer Communication Review*, 38(2):69–74, 2008.
66. Nick McKeown, Tom Anderson, Hari Balakrishnan, Guru Parulkar, Larry Peterson, Jennifer Rexford, Scott Shenker, and Jonathan Turner. OpenFlow: Enabling Innovation in Campus Networks. *ACM SIGCOMM Computer Communication Review*, 38(2):69–74, 2008.
67. Omar Mehanna, Nicholas D Sidiropoulos, and Georgios B Giannakis. Joint Multicast Beamforming and Antenna Selection. *IEEE Transactions on Signal Processing*, 61(10):2660–2674, 2013.
68. Sevil Mehraghdam, Matthias Keller, and Holger Karl. Specifying and Placing Chains of Virtual Network Functions. In *Proceedings of the 3rd International Conference on Cloud Networking (CloudNet)*, pages 7–13. IEEE, 2014.
69. Albert Mestres, Alberto Rodriguez-Natal, Josep Carner, Pere Barlet-Ros, Eduard Alarcón, Marc Solé, Victor Muntés-Mulero, David Meyer, Sharon Barkai, Mike J Hibbett, et al. Knowledge-defined networking. *ACM SIGCOMM Computer Communication Review*, 47(3):2–10, 2017.
70. Toshiaki Miyazaki. Dynamic Function Alternation to Realize Robust Wireless Sensor Network. *International Journal of Handheld Computing Research (IJHCR)*, 3(3):17–34, July 2012.
71. Toshiaki Miyazaki, Daiki Shitara, Yuji Endo, Yuuki Tanno, Hidenori Igari, and Ryouhei Kawano. Die-hard Sensor Network: Robust Wireless Sensor Network Dedicated to Disaster Monitoring. In *Proceedings of the 5th International Conference on Ubiquitous Information Management and Communication (ICUIMC)*, pages 53:1–53:10. ACM, 2011.
72. China Mobile. C-RAN: the Road Towards Green RAN. *White Paper, ver*, 2, 2011.
73. Hendrik Moens and Filip De Turck. VNF-P: A Model for Efficient Placement of Virtualized Network Functions. In *Proceedings of the 10th International Conference on Network and Service Management (CNSM)*, pages 418–423. IEEE, 2014.

74. Irakli Nadareishvili, Ronnie Mitra, Matt McLarty, and Mike Amundsen. *Microservice architecture: aligning principles, practices, and culture.* "O'Reilly Media, Inc.", 2016.

75. Akihiro Nakao. Software-defined data plane enhancing sdn and nfv. *IEICE Transactions on Communications*, 98(1):12–19, 2015.

76. Michael J. Neely. Opportunistic scheduling with worst case delay guarantees in single and multi-hop networks. In *2011 Proceedings IEEE INFOCOM*, pages 1728–1736. IEEE, apr 2011.

77. Felicián Németh, Balázs Sonkoly, Levente Csikor, and András Gulyás. A large-scale multipath playground for experimenters and early adopters. In *ACM SIGCOMM Computer Communication Review*, volume 43, pages 481–482. ACM, 2013.

78. Kien Nguyen, Quang Tran Minh, and Shigeki Yamada. A Software-Defined Networking Approach for Disaster-Resilient WANs. In *22nd International Conference on Computer Communications and Networks (ICCCN)*, pages 1–5. IEEE, 2013.

79. Yipei Niu, Bin Luo, Fangming Liu, Jiangchuan Liu, and Bo Li. When hybrid cloud meets flash crowd: Towards cost-effective service provisioning. In *2015 IEEE Conference on Computer Communications (INFOCOM)*, pages 1044–1052. IEEE, apr 2015.

80. Dusit Niyato, Xiao Lu, and Ping Wang. Adaptive Power Management for Wireless Base Stations in a Smart Grid Environment. *IEEE Wireless Communications*, 19(6), 2012.

81. Charles Pandana and KJ Ray Liu. Near-optimal reinforcement learning framework for energy-aware sensor communications. *IEEE Journal on Selected Areas in Communications*, 23(4):788–797, 2005.

82. Milan Patel, Brian Naughton, Caroline Chan, Nurit Sprecher, Sadayuki Abeta, Adrian Neal, et al. Mobile-edge computing introductory technical white paper. *White paper, mobile-edge computing (MEC) industry initiative*, pages 1089–7801, 2014.

83. Cesare Pautasso, Olaf Zimmermann, Mike Amundsen, James Lewis, and Nicolai Josuttis. Microservices in practice, part 2: Service integration and sustainability. *IEEE Software*, (2):97–104, 2017.

84. Ken Pepple. *Deploying openstack.* "O'Reilly Media, Inc.", 2011.

85. M L Puterman. *Markov decision processes: Discrete stochastic dynamic programming.* John Wiley & Sons, Inc., 1994.

86. Yinan Qi, Muhammad Z Shakir, Muhammad A Imran, Atta Quddus, and Rahim Tafazolli. How to Solve the Fronthaul Traffic Congestion Problem in H-CRAN? In *Communications Workshops (ICC), 2016 IEEE International Conference on*, pages 240–245. IEEE, 2016.

87. Christian Robert. *Machine Learning, a Probabilistic Perspective.* MIT Press, 2012.

88. Tiago Gama Rodrigues, Katsuya Suto, Hiroki Nishiyama, and Nei Kato. Hybrid Method for Minimizing Service Delay in Edge Cloud Computing Through VM Migration and Transmission Power Control. *IEEE Transactions on Computers*, 66(5):810–819, 2017.

89. Tiago Gama Rodrigues, Katsuya Suto, Hiroki Nishiyama, Nei Kato, and Katsuhiro Temma. Cloudlets Activation Scheme for Scalable Mobile Edge Computing with Transmission Power Control and Virtual Machine Migration. *IEEE Transactions on Computers*, 2018.

90. Mathew Ryden, Kwangsung Oh, Abhishek Chandra, and Jon Weissman. Nebula: Distributed Edge Cloud for Data Intensive Computing. In *2014 IEEE International Conference on Cloud Engineering*, pages 57–66. IEEE, mar 2014.

91. Dario Sabella, Antonio De Domenico, Efstathios Katranaras, Muhammad Ali Imran, Marco Di Girolamo, Umer Salim, Massinissa Lalam, Konstantinos Samdanis, and Andreas Maeder. Energy Efficiency Benefits of RAN-as-a-Service Concept for a Cloud-based 5G Mobile Network Infrastructure. *IEEE Access*, 2:1586–1597, 2014.

92. Jaspal S Sandhu, Alice M Agogino, Adrian K Agogino, et al. Wireless sensor networks for commercial lighting control: Decision making with multi-agent systems. In *AAAI workshop on sensor networks*, volume 10, pages 131–140, 2004.

93. Gregor Schaffrath, Christoph Werle, Panagiotis Papadimitriou, Anja Feldmann, Roland Bless, Adam Greenhalgh, Andreas Wundsam, Mario Kind, Olaf Maennel, and Laurent Mathy. Network virtualization architecture: Proposal and initial prototype. In *Proceedings of the 1st ACM workshop on Virtualized infrastructure systems and architectures*, pages 63–72. ACM, 2009.

94. Stefan Schmid and Jukka Suomela. Exploiting Locality in Distributed SDN Control. In *Proceedings of the Second ACM SIGCOMM Workshop on Hot Topics in Software Defined Networking*, pages 121–126. ACM, 2013.

95. Sandra Scott-Hayward, Gemma O'Callaghan, and Sakir Sezer. Sdn security: A survey. In *2013 IEEE SDN For Future Networks and Services (SDN4FNS)*, pages 1–7. IEEE, 2013.

96. Danbing Seto, John P Lehoczky, Lui Sha, and Kang G Shin. On task schedulability in real-time control systems. In *Proceedings of the 17th Real-Time Systems Symposium (RTSS)*, pages 13–21. IEEE, 1996.

97. Sakir Sezer, Sandra Scott-Hayward, Pushpinder Kaur Chouhan, Barbara Fraser, David Lake, Jim Finnegan, Niel Viljoen, Marc Miller, and Navneet Rao. Are we ready for sdn? implementation challenges for software-defined networks. *IEEE Communications Magazine*, 51(7):36–43, 2013.

98. Justine Sherry, Shaddi Hasan, Colin Scott, Arvind Krishnamurthy, Sylvia Ratnasamy, and Vyas Sekar. Making middleboxes someone else's problem: network processing as a cloud service. *ACM SIGCOMM Computer Communication Review*, 42(4):13–24, 2012.

99. W. Shi, J. Cao, Q. Zhang, Y. Li, and L. Xu. Edge computing: Vision and challenges. *IEEE Internet of Things Journal*, 3(5):637–646, Oct 2016.

100. Yuanming Shi, Jun Zhang, and Khaled B Letaief. Group Sparse Beamforming for Green Cloud-RAN. *IEEE Transactions on Wireless Communications*, 13(5):2809–2823, 2014.

101. Yuanming Shi, Jun Zhang, Brendan O'Donoghue, and Khaled B Letaief. Large-scale Convex Optimization for Dense Wireless Cooperative Networks. *IEEE Transactions on Signal Processing*, 63(18):4729–4743, 2015.

102. William Stallings. *Foundations of modern networking: SDN, NFV, QoE, IoT, and Cloud*. Addison-Wesley Professional, 2015.

103. Brent Stephens, Alan Cox, Wes Felter, Colin Dixon, and John Carter. PAST: Scalable Ethernet for data centers. In *Proceedings of the 8th international conference on Emerging networking experiments and technologies*, pages 49–60. ACM, 2012.

104. Yi Sun, Xiaoqi Yin, Junchen Jiang, Vyas Sekar, Fuyuan Lin, Nanshu Wang, Tao Liu, and Bruno Sinopoli. Cs2p: Improving video bitrate selection and adaptation with data-driven throughput prediction. In *Proceedings of the 2016 ACM SIGCOMM Conference*, pages 272–285. ACM, 2016.

105. Katsuya Suto, Hiroki Nishiyama, and Nei Kato. Postdisaster User Location Maneuvering Method for Improving the QoE Guaranteed Service Time in Energy Harvesting Small Cell Networks. *IEEE Transactions on Vehicular Technology*, 66(10):9410–9420, 2017.

106. Fengxiao Tang, Zubair Md Fadlullah, Nei Kato, Fumie Ono, and Ryu Miura. AC-POCA: Anticoordination Game Based Partially Overlapping Channels Assignment in Combined UAV and D2D-Based Networks. *IEEE Transactions on Vehicular Technology*, 67(2):1672–1683, 2018.

107. Fengxiao Tang, Zubair Md Fadlullah, Bomin Mao, Nei Kato, Fumie Ono, and Ryu Miura. On A Novel Adaptive UAV-Mounted Cloudlet-Aided Recommendation System for LBSNs. *IEEE Transactions on Emerging Topics in Computing*, 2018.

108. Sibel Tombaz, Paolo Monti, Kun Wang, Anders Vastberg, Marco Forzati, and Jens Zander. Impact of Backhauling Power Consumption on the Deployment of Heterogeneous Mobile Networks. In *Global Telecommunications Conference (GLOBECOM 2011), 2011 IEEE*, pages 1–5. IEEE, 2011.

109. Amin Tootoonchian and Yashar Ganjali. HyperFlow: A Distributed Control Plane for OpenFlow. In *Proceedings of the Internet Network Management Conference on Research on Enterprise Networking*, pages 3–3. USENIX Association, 2010.

110. Amin Tootoonchian, Sergey Gorbunov, Yashar Ganjali, Martin Casado, and Rob Sherwood. On Controller Performance in Software-Defined Networks. In *USENIX Workshop on Hot Topics in Management of Internet, Cloud, and Enterprise Networks and Services (Hot-ICE)*, volume 54, 2012.

111. Laz Vekiarides. 5 bitter truths about software-defined storage. https://www.infoworld.com/article/2997239/5-bitter-truths-about-software-defined-storage.html/, 2015. [Online; accessed 15-July-2019].

112. Bang Wang. Coverage Problems in Sensor Networks: A Survey. *ACM Computer Survey*, 43(4):32:1–32:53, Oct. 2011.
113. Jie Wang and Ning Zhong. Efficient point coverage in wireless sensor networks. *Journal of Combinatorial Optimization*, 11(3):291–304, 2006.
114. Qixing Wang, Dajie Jiang, Jing Jin, Guangyi Liu, Zhigang Yan, and Dacheng Yang. Application of BBU+ RRU based CoMP system to LTE-Advanced. In *Communications Workshops, 2009. ICC Workshops 2009. IEEE International Conference on*, pages 1–5. IEEE, 2009.
115. Zhaoguang Wang, Zhiyun Qian, Qiang Xu, Zhuoqing Mao, and Ming Zhang. An untold story of middleboxes in cellular networks. In *ACM SIGCOMM Computer Communication Review*, volume 41, pages 374–385. ACM, 2011.
116. Ami Wiesel, Yonina C Eldar, and Shlomo Shamai. Linear Precoding via Conic Optimization for Fixed MIMO Receivers. *IEEE Transactions on Signal Processing*, 54(1):161–176, 2006.
117. Rebecca Willett, Aline Martin, and Robert Nowak. Backcasting: adaptive sampling for sensor networks. In *Proceedings of the 3rd international symposium on Information processing in sensor networks*, pages 124–133. ACM, 2004.
118. Nick Wingfield. Amazon's profits grow more than 800 percent, lifted by cloud services. *New York Times*, 28, 2016.
119. Zichuan Xu, Weifa Liang, Meitian Huang, Mike Jia, Song Guo, and Alex Galis. Approximation and Online Algorithms for NFV-Enabled Multicasting in SDNs. In *IEEE International Conference on Distributed Computing Systems*, pages 625–634, 2017.
120. Navindra Yadav, Jim Guichard, Brad McConnell, Christian Jacquenet, Michael Smith, Abhishek Chauhan, Mohamed Boucadair, Paul Quinn, Rajeev Manur, Tom Nadeau, and Others. Network Service Chaining Problem Statement. *Network*, 2013.
121. Long Yao, Peilin Hong, and Wei Zhou. Evaluating the Controller Capacity in Software Defined Networking. In *23rd International Conference on Computer Communication and Networks (ICCCN), 2014*, pages 1–6, Aug 2014.
122. David K Y Yau, Nung Kwan Yip, Chris Y T Ma, Nageswara S V Rao, and Mallikarjun Shankar. Quality of monitoring of stochastic events by periodic and proportional-share scheduling of sensor coverage. *ACM Transactions on Sensor Networks (TOSN)*, 7(2):18, 2010.
123. Mohamed Younis, Moustafa Youssef, and Khaled Arisha. Energy-aware routing in cluster-based sensor networks. In *Modeling, Analysis and Simulation of Computer and Telecommunications Systems, 2002. MASCOTS 2002. Proceedings. 10th IEEE International Symposium on*, pages 129–136. IEEE, 2002.
124. Yuan Yuan, Paramvir Bahl, Ranveer Chandra, Thomas Moscibroda, and Yunnan Wu. Allocating dynamic time-spectrum blocks in cognitive radio networks. In *Proceedings of the 8th ACM international symposium on Mobile ad hoc networking and computing*, pages 130–139. ACM, 2007.
125. Frank Yue. Network functions virtualization-everything old is new again. *F5 Neworks*, 2013.
126. Deze Zeng, Lin Gu, Song Guo, Zixue Cheng, and Shui Yu. Joint Optimization of Task Scheduling and Image Placement in Fog Computing Supported Software-Defined Embedded System. *IEEE Transactions on Computers*, 65(12):3702–3712, 2016.
127. Deze Zeng, Jie Zhang, Song Guo, Lin Gu, and Kun Wang. Take Renewable Energy into CRAN toward Green Wireless Access Networks. *IEEE Network*, 31(4):62–68, 2017.
128. Qi Zhang, Lu Cheng, and Raouf Boutaba. Cloud computing: state-of-the-art and research challenges. *Journal of internet services and applications*, 1(1):7–18, 2010.
129. Jian Zhao, Tony QS Quek, and Zhongding Lei. Coordinated Multipoint Transmission with Limited Backhaul Data Transfer. *IEEE Transactions on Wireless Communications*, 12(6):2762–2775, 2013.
130. Xiaorong Zhu, Lianfeng Shen, and Tak-Shing Peter Yum. Analysis of cognitive radio spectrum access with optimal channel reservation. *IEEE Communications Letters*, 11(4):304–306, 2007.
131. L. Gkatzikis, G. Iosifidis, I. Koutsopoulos, and L. Tassiulas. Collaborative placement and sharing of storage resources in the Smart Grid, in *2014 IEEE International Conference on Smart Grid Communications, SmartGridComm 2014*, 2015, pp. 103–108.

Printed in the United States
By Bookmasters